DESCENT FROM HEAVEN

American Philological Association American Classical Studies

Susan Treggiari, editor

The Harmonics of Nicomachus and the Pythagorean Tradition	Flora R. Levin
The Etymology and the Usage of ΠΕΙΡΑΡ *in Early Greek Poetry*	Ann L. T. Bergren
Two Studies in Roman Nomenclature	D. R. Shackleton Bailey
The Latin Particle Quidem	J. Solodow
On the Hymn to Zeus in Aeschylus' Agamemnon	Peter M. Smith
The Andromache *of Euripides*	Paul David Kovacs
A Commentary on the Vita Hadriani *in the* Historia Augusta	Herbert W. Benario
Creation and Salvation in Ancient Orphism	Larry J. Alderink
Eros Sophistes: *Ancient Novelists at Play*	Graham Anderson
Ancient Philosophy and Grammar: The Syntax *of Apollonius Dyscolus*	David Blank
Autonomia: *Its Genesis and Early History*	Martin Ostwald
Language and Metre: Resolution, Porson's Bridge, and Their Prosodic Basis	A. M. Devine L. D. Stephens
Descent from Heaven: Images of Dew in Greek Poetry and Religion	Deborah Boedeker

Deborah Boedeker

Descent from Heaven

Images of Dew in Greek Poetry and Religion

Scholars Press
Chico, California

DESCENT FROM HEAVEN
Images of Dew in Greek Poetry
and Religion

Deborah Boedeker

Library of Congress Cataloging in Publication Data

Boedeker, Deborah Dickmann.
Descent from heaven.

(American classical studies ; no. 13)
Bibliography: p.
Includes index.
1. Greek poetry—History and criticism. 2. Dew in literature. 3. Greece—Religion. 4. Dew—Religious aspects. 5. Eérsē (The Greek word) 6. Drósos (The Greek word) I. Title. II. Series
PA3015.N4D483 1984 881'.01'0936 84-20292
ISBN 0-89130-807-5

Printed in the United States of America
on acid-free paper

Acknowledgments

This essay has been long in the making, and has profited along the way from the advice and support of many friends and colleagues. Parts of Chapters Three, Four and Five were presented in colloquia at the Center for Hellenic Studies, Princeton, Brown, and Holy Cross. The ensuing discussions produced many useful suggestions, most of which I have incorporated into the final product. I am especially grateful to Froma Zeitlin, Alison Elliott, and William J. Ziobro for their interest and practical advice, and most of all to Ann Bergren, who encouraged me to develop my work on dew into a monograph. Early drafts of the manuscript were read by Mary Lefkowitz and Charles Segal, who generously offered a number of incisive comments. Gregory Nagy read a later version; for his learned, frank, and friendly criticism I am again gratefully indebted to him.

Thanks are also due the Editorial Board for Monographs of the American Philological Association, especially chairpersons Susan Treggiari and Deborah Hobson, for overseeing the development of this study to its final form. Two anonymous referees for the Board provided careful readings, critical in the best sense of the term, which have helped me to correct many mistakes and clarify many points. In the final stages of editing this text I have been fortunate to have the assistance of Ruthann Whitten, Elli Mylonas, and Dick Damon, whose good cheer and expertise have greatly lightened the burden of preparing camera-ready copy. Less personal perhaps but no less important has been the financial support awarded by the College of the Holy Cross, which has helped greatly with the expenses of research and editing.

Finally I am happy to record my deepest thanks to my children, Edgar and Nancy, for their patient support, and to my husband Kurt Raaflaub for his invaluable encouragement and help.

Deborah Boedeker
Providence, August 1984

CONTENTS

Introduction

THE "ARCHAEOLOGY" OF DEW:

BACKGROUND AND METHODS

Ancient Greek has two words which we usually translate as 'dew': *eérsē* (with its variants) and *drósos*.[1] This essay will examine these and related words in their contexts in epic, lyric, and tragedy in order to suggest what a contemporary audience, attuned to the nuances of traditional Greek poetic diction, would have understood in their use. Words for 'dew' occur in contexts involving some of the major concerns of Greek culture--male and female roles in procreation, communication between gods and mortals, human work and its relation to the realm of nature, the contrast between life and death, the nature and functions of language. From our cultural perspective it is scarcely apparent what 'dew' has to do with such issues, or what they may have in common with each other that causes them to elicit the same image. A study of the semantics of *eérsē* and *drósos* may help to answer these questions. But in addition to examining how

----

1 Occasionally I use *'dew'* enclosed in single quotes to refer to *eérsē* or *drósos* without reference to their actual significance in context--so for example even in passages where *drósos* refers to river or spring water. Although it is clear that *eérsē* and *drósos* are not a perfect semantic match for English 'dew', I have adopted this convention not only as a means to avoid tedious repetition of the Greek words, but also because in normal, unheightened language both words clearly denote what we call 'dew' (moisture condensed from the air, especially at night, resulting from a decrease in temperature: cf. the uses of *drósos* in prose). Where I write *dew* without quotes I mean the substance in the literal sense.

the image of 'dew' is used in early Greek poetry, it is hoped that this essay may also illuminate some of the underlying concerns and assumptions in the passages where 'dew' appears.

Both *eérsē* and *drósos* will be examined with reference to the genre, author, and text in which they occur, but most importantly in terms of their immediate semantic context. Words signifying other kinds of water, especially water from the sky, will also be considered where relevant, in order to determine their convergences and divergences with *eérsē* and *drósos* and thereby to clarify the specific connotations of words for 'dew'.

We can begin by surveying the linguistic history of Homeric *eérsē*, whose etymology is well established, although not without certain problems of detail.[2] A thematic noun from the root **wers-*, it is cognate with Indic *varṣám* 'rain' and Hittite *waršaš* 'dew', and related to Irish *frass* 'rain'. In Greek it is related to *ouranós* 'sky' (as source of rain?)[3] and, more surprisingly, to *ouréō* 'urinate'--a fact which recalls the occasional analogies between rain and urine (and male fecundity, an idea whose importance in this connection will be discussed later) in early Indic poetry.[4]

----

[2] See Appendix I: The Variants of *Eérsē*.

[3] The shift from ἐρσ- to οὐρ- is explained by Wackernagel 1888, refined by Miller 1976 esp. p. 159, as a result of the regular alternation between voiced and voiceless /s/ in relation to the location of the word-accent; hence ἔρση (accent preceding) but οὐρέω (accent following).

In addition to words meaning 'rain' or 'dew' and the like, **wers-* also produces derivatives meaning 'male animal': Indic *vṛṣa-bhá* 'bull, male', Latin *verres* 'boar', etc. Such derivatives suggest that there may be a correspondence between **wers-* and the rhyming root **ers-*, which generates Greek *érsēn* / *ársēn* 'male', Indic *ṛṣa-bhá* 'bull, male', etc., but also Latin *ros* 'dew, pure water'. Benveniste maintains that the two roots are separate except in Indic, where their convergence is a secondary development not found in other languages.[5] He defines **ers-* as signifying the *physical* male, while **wers-* sometimes comes to mean the *functional*, fertilizing male (*reproducteur*) by an extension of its primary significance 'liquid distilled from the sky', often attested in the specialized sense of 'rain as fecundating moisture'. Benveniste is convincing with regard to the primary and secondary meanings of **wers-* suggested here, yet the correspondences between **ers-* and **wers-* are too close to be explained as essentially unrelated rhyme-forms, especially since other such pairs with and without initial /w/ can be cited for Indo-European, as Watkins and Nagy have shown.[6]

The two roots **wers-* 'rain, fecundating moisture' and **ers-* 'male' then

----

4 See Nagy 1974.231-237 for analogies between rain and urine with reference to the Maruts in the *Rig Veda*, esp. in *RV* 1.64.6, 2.34.13, and 5.58.7.

5 Benveniste 1969 vol. 1.21-25. So too Chantraine 1968-77, *s.v.* ἕρση.

6 Watkins 1973.88-89; so too Nagy 1974a 72-73, except that Nagy posits an initial laryngeal for this root, so that the alternation of rhyme forms is in the pattern CweC(C)-/CeC(C), i.e. $*H^1$*wers-*/H^1*ers-*.

resemble each other in meaning as well as in morphology. This correspondence is suggestive for the present study; it has implications not only for the Greek uses of 'dew,'[7] but also for understanding a particular set of contexts in which *eérsē* and its relative *ouranós* sometimes appear: the Indo-European concept of male as *reproducteur*. In this perspective fluid from a male, especially from a male divinity associated with rain or the sky, may be generative in and of itself. Hence in the *Rig Veda* the Maruts, gods associated with rain clouds, are said to "deposit their strength (in the earth) as a husband deposits the embryo" (*RV* 5.58.7)--and the "strength" of the Maruts is precisely the rain (*varṣám*) they send. As we shall see, this complex of ideas is reflected in Greek cosmology, where Zeus or Ouranos is a fecundating god *par excellence*; in natural history, where 'dew' can generate life spontaneously; and in a more general sense in certain Greek views on procreation, in which the male role is pre-eminent, while the female serves merely as a temporary vessel for the generative seed, until the embryo emerges as the offspring of its father.

The generative significance of *_wers_- is reflected to a large extent in the semantic range of its derivatives *eérsē* and *ouranós* (when personified as a god of the sky) in early Greek poetry. By contrast, when other words for 'liquid from the sky' (e.g. *ómbros* 'rainstorm' or *húetos* 'rain') occur in a context with-

----

7 This applies not only to *eérsē* but to *drósos* as well, whether because it signifies the same kind of 'moisture' as *eérsē* or because it has an etymological relationship to the reflexes of *_wers_- and *_ers_-. See Appendix II for discussion of the possible derivation of *drósos* from the root *_ers_-.

out one of the **wers-* derivatives, the moisture is almost always destructive or unpleasant rather than fertilizing. Even in Greek prose, where *drósos* is usually attested, a derivative of *eérsē* (the adjective *hersṓdēs*), describes air which is not only moist but explicitly generative (Theophrastus, *C.P.* 3.2.6).

The normal term for dew in prose, and generally in literature of the fifth century and later, is *drósos*, whose connotations closely resemble those of *eérsē*. Although the noun itself is not attested until Pindar, Aeschylus, and Herodotus, adjectives apparently derived from *drósos* appear twice in the poems of Sappho, where *éersa* is also attested. In one of these attestations, fr. 95.12 LP, *drosóentas* 'dewy' is used together with *lōtínois* 'lotus-y', to describe the banks of the underworld river Acheron. In the other, fr. 71.8 LP, *drosóessa* 'dewy' describes something (probably *aēdṓn* 'singer) in a context of sweet song. It is significant that the semantic range of *dros-* in Sappho corresponds precisely to uses of *eérsē* or its derivatives in epic (lotus: *Il.* 14.348, of the 'dewy lotus' produced by Earth; song: *Sh. Her.* 395, of the cicada who eats dew and sings in the summer heat). In Pindar, who is the only early poet besides Sappho to use both *éers-* and *drós-*, the semantic ranges of the two words likewise seem to converge. Later poets, however, especially Euripides, use *drósos* with a broader range of meanings. Whereas *eérsē* usually means literal dew (or possibly rain-drops), *drósos* can refer to sea, river, or spring water as well. In tragedy, in fact, it occasionally signifies water from a source other than the sky (usually qualified by an adjective meaning 'of the sea', 'of a river', etc.). Yet even in these extended uses of the word, it retains much of its marked significance, suggesting qualities of fertility or purity that derive from the celestial and cosmological associations of *eérsē*.

The semantic correspondence of *drósos* to *eérsē* is so close, and its phonological resemblance to Latin *ros* (from the rhyming root **ers-*) so great, that it is tempting to postulate a closer identity between the two Greek words, as Meillet has done by suggesting that *drósos* actually derives from **ers-*.[8] Be that as it may, the main point is that the semantics of *drósos* closely resemble those of *eérsē*, whatever the resemblance in their morphology may be, and it is primarily questions of semantics which this essay will address.

First, we will analyze the uses of the **wers-* derivatives *eérsē* and *ouranós*, as well as of *drósos*, in contexts where generativity (always from a male or from the sky) is implied. Next the words will be examined in passages that deal with growth and other forms of transformation--an extension, perhaps, of the primary idea of fertilizing moisture. The fertility attributed to 'dew' in Greek poetry will be compared with the use of Greek *drósos* and Latin *ros* in agricultural writings, with some surprising results: from the standpoint of practical agriculture, dew can connote a set of experiences very different from its usual associations in poetry. Contexts of generativity and fertility comprise the most common semantic range of 'dew' in Greek poetry, and this range of meanings seems to underlie other passages where 'dew' is used in an ironic or paradoxical

----

8 Meillet 1931.234-6. See Appendix II: An Etymology of *Drósos?*

sense, where, for example, it occurs as a feature of rather ambivalent erotic settings or even as a metaphor for blood. Several such "developed" uses of *drósos* and *eérsē* will be the subject of another chapter. Then the broad and interesting set of analogies between 'dew' and sound--particularly language--will be considered in some detail. Finally, the significance of 'dew' in Greek poetry will lead us to the Athenian Acropolis, in the context of the birth of Erichthonius from the 'dew' of Hephaestus and his fosterage at the hands of Herse, Pandrosus, and Aglaurus, the three daughters of Cecrops whom modern scholarship knows as the "Dew-Sisters". In addition to the myth of Erichthonius, and perhaps related to it, is the Athenian ritual known as Hersephoria or Arrhephoria, which was performed by girls dedicated to Pandrosus and which involved their descent to a shrine of Aglaurus. The name of this rite has been interpreted by some ancient and modern scholars as 'Dew-Bearing' (with the element *hersē-* or *arrē-* from **wers-*). What we shall learn about 'dew' in Greek culture, and particularly in Athenian literature, will help us evaluate this etymology and analyze the possible role of 'dew' in the mysterious ritual.

In sum, this essay attempts an "archaeology of imagination",[9] or more modestly, the archaeology of an image. Its aim is always to see the image of 'dew' in context and to deduce what kinds of associations it brings to that context. Of necessity, the method involves a great deal of cross-referencing; pas-

----

9 Landow 1982.15.

sages are cited with respect to one range of meaning and then recur in another, casting (it is hoped) some light on a context by recalling the significance of 'dew' from another perspective. In particular, the "cosmological" perspective of 'dew'--which closely reflects the richly connotative etymology of *éersē*--is frequently brought to mind.

Although the main focus of this essay is early Greek poetry--traditional epic, elegy, lyric, and their close relative Attic tragedy--I have not limited myself entirely to uses of 'dew' in these genres. The reason for this is twofold. First, the attributes of dew in other genres, for example in the natural history of Aristotle or the scientific and literary comments of Plutarch, can often illuminate notions associated with 'dew' in early poetry, by indicating ideas connected with the substance in common observation and folklore. And second, the differences among perceptions of dew in different genres can itself be illuminating, by showing us what is evoked by 'dew' in traditional poetry, sometimes in contrast to the ways in which it was perceived on a more naturalistic level. I cite some Latin authors as well as Greek for this purpose, especially Columella and Pliny the Elder. It may be argued that such sources are not relevant to this study, since the culture of Rome in the early Empire differs so greatly from that of archaic and classical Greece. But despite the vast cultural differences, I believe that these authors can still be useful in a discussion of dew in Greek culture, either because like Columella they present a non-poetic, pragmatic perspective on dew which may serve as a kind of "control" on the viewpoint of archaic Greek poetry, or because like Pliny, dependent on Greek sources of natural history, they may more fully develop some of the popular conceptions of the Greeks who are our primary objects of study, and thereby supply illuminating supplementary information. In either case, I find their information worth comparing with the views of 'dew' attested in Greek poetry.

The result of this study is to suggest a few specific ways in which traditional Greek poetry reflects a world-view very different from our own, different too from the more secular or scientific world-views of later centuries in antiquity, even though it was influential in shaping their assumptions (e.g. about the role of dew in spontaneous generation). In the archaic Greek world-view, the natural world coalesces with the divine on one hand and the human on the other, sometimes providing a very direct medium through which the two may communicate with each other. Thus moisture from the sky may appear as a sign from Zeus--a portent of calamity in warfare (*Il.* 11.52-55), a punishment for the evil deeds of men (*Il.* 16.385)--or as the fructifying moisture that falls on grasses and flowers, recalling the ease and abundance of the Golden Age before mortals had to earn their bread and wine by cultivating fields and vines. 'Dew' from the sky, above all, serves as a reminder to human beings of the happy fertility and ethereal purity that lies, too often, just beyond their reach.

Chapter One

GENERATION AND BIRTH

Writing about Semitic and Indo-European world-views, the folklorist Alan Dundes has observed:

> Life depends on liquid. From the concept of the "water of life" to semen, milk, blood, bile, saliva, and the like, the consistent principle is that liquid means life while loss of liquid means death.[1]

Fertility, generativity, and vitality, associated with bodily fluids on the individual human level, are attributed to the waters of springs, rivers, the sea, rain, and so on in the broader perspective of the natural world. In his study of ancient beliefs about water, M. Ninck discusses the generative properties of water, directing his attention first to *aqua viva* or chthonic water.[2] As Ninck observes, in Greek terms relatively still waters such as springs are considered feminine; flowing, turbulent waters such as rivers are masculine. In myths, correspondingly, springs may originate in the metamorphosis of a female such as Dirce at Thebes.[3] Rivers, of course, are personified as males--bulls or horned anthropomorphic gods--and many myths emphasize the sexual nature of a river god or of Poseidon, the Olympian god of waters, appearing in a river or the

----

[1] Dundes 1980.101. Cf. also Onians 1951.190-193 and 200-228.

[2] Ninck 1921.10-31 and *passim.*

[3] Ninck p. 11.

sea.[4]

Male generative power then is associated with flowing waters on the earth. A similar power is also attributed to the waters of the sky, again from the perspective of natural process (which I shall refer to as the "cosmological" level) and in terms of myth as well. In this system the **wers*-derivative *Ouranós* 'sky' (related to *eérsē* 'dew'), figures prominently. In Greek as in other cultures, the observation that moisture from the sky encourages the growth of plants is often expressed in anthropomorphic terms, in which a male Sky impregnates a female Earth with rain. One such description is the well-known fragment of Aeschylus' *Danaids*, quoted in Athenaeus as an argument for the power of love (Aesch. fr. 44N: Athenaeus 13.600a):[5]

> ἐρᾶι μὲν ἁγνὸς Οὐρανὸς τρῶσαι χθόνα,
> ἔρως δὲ Γαῖαν λαμβάνει γάμου τυχεῖν,
> ὄμβρος δ' ἀπ' εὐνάεντος Οὐρανοῦ πεσὼν
> ἔκυσε Γαῖαν, ἡ δὲ τίκτεται βροτοῖς
> μήλων τε βοσκὰς καὶ βίον Δημήτριον,
> δενδρῶτις ὥρα δ' ἐκ νοτίζοντος γάμου
> τέλειός ἐστι· τῶν δ' ἐγὼ παραίτιος.

> Holy Ouranos desires to pierce the earth
> and desire holds Gaia to share in marriage;
> rain (*ómbros*) falling from fair-flowing Ouranos
> impregnates Gaia, and she brings forth for mortals

----

4 Ninck, p. 14, 25-27.

5 Ninck p. 26. See also Motte 1973. 219-221 for a discussion of this fragment and of other passages, including several in Modern Greek, which refer to the marriage of earth and sky. Motte's reading of the Aeschylus fragment differs from mine in several respects, especially in his translation of εὐνάεντος as *époux* (from εὐνή), rather than 'fair-flowing' (from νάω).

flocks of sheep and Demeter's source of life.
From the moist marriage the ripeness of trees
is fulfilled. Of these things I am the cause.

These words are attributed to Aphrodite herself. Similar in thought and diction is the passage from an unknown tragedy of Euripides quoted in the same context of Athenaeus (Eur. fr. 298.7-13N: Athenaeus 13.599f-600a):

ἐρᾶι μὲν ὄμβρου Γαῖ', ὅταν ξηρὸν πέδον
ἄκαρπον αὐχμῶι νοτίδος ἐνδεῶς ἔχηι
ἐρᾶι δ' ὁ σεμνὸς Οὐρανὸς πληρούμενος
ὄμβρου πεσεῖν εἰς Γαῖαν Ἀφροδίτης ὕπο·
ὅταν δὲ συμμιχθῆτον ἐς ταὐτὸν δύο,
φύουσιν ἡμῖν πάντα καὶ τρέφουσ' ἅμα,
δι' ὧν βρότειον ζῆι τε καὶ θάλλει γένος.

Gaia desires rain whenever her dry plain,
fruitless with drought, needs moisture;
and revered Ouranos desires, when he is filled
with rain, to fall onto Gaia because of Aphrodite.
Whenever these two mingle into one
they engender and nurture for us all things
through which the mortal race lives and thrives.

The first of these fragments is spoken by Aphrodite herself; the second follows and illustrates an affirmation of that goddess' supreme power (Eur. fr. 298.1-6N). In both fragments, the model of Sky and Earth, Ouranos and Gaia, is used to prove to a human audience the necessity and primacy of sexual love. Submission to Eros and Aphrodite is implicitly justified by the cosmological parallel, in which all life on earth is generated and increased by the "moist marriage" of the primal "parents". The anthropomorphic model underlying the relationship of Sky and Earth is thus brought full circle, as the cosmological pattern comes to serve as model for human behavior. A passage in the Hesiodic *Works and Days* provides a striking "naturalistic" parallel to the anthropomorphic descriptions of Ouranos fecundating Gaia (*WD* 548-549):

> ἠῶος δ' ἐπὶ γαῖαν ἀπ' οὐρανοῦ ἀστερόεντος
> ἀὴρ πυρόφορος τέταται μακάρων ἐπὶ ἔργοις
>
> To earth from starry *ouranós* at dawn
> falls wheat-bringing *aḗr* on the fields of the blessed.

Here the fertilizing[6] *aḗr* 'mist, moist air' that falls at dawn resembles dew more closely than rain. Gentle moisture from sky or air is again closely associated with vegetal fertility in an Aristotelian treatise and in Theophrastus, in passages discussed at the end of this chapter.

In Latin poetry also the union of Earth and Sky through rain is seen as the source of life on earth, as when Lucretius describes their marriage in his argument for the imperishability of matter (Lucr. 1.250-251):[7]

> postremo pereunt imbres, ubi eos pater aether
> in gremium matris terrai praecipitavit
>
> finally rain-drops disappear, when Father Aether
> has dropped them into the lap of Mother Earth . . .

The passage continues by describing the plant, animal, and human life engendered and nurtured by the heavenly moisture.

In a Vergilian description evidently influenced by the Lucretian lines, the

----

6 *Pace* West 1978.296, who questions the mss. reading πυρόφορος in line 548, and prefers the reading πυροφόροις to go with the dative ἔργοις 'wheat-bearing fields'.

7 See also Lucr. 2.991-998.

emphasis in on earth's need for fertilizing rains (*G.* 2.324-327):[8]

> vere tumunt terrae et genitalia semina poscunt.
> tum pater omnipotens fecundis imbribus aether
> coniugis in gremium laetae descendit, et omnes
> magnus alit magno commixtus corpore fetus.
>
> In spring lands swell and ask for fertilizing seeds.
> Then almighty Father Aether falls with fertile rains
> into his happy spouse's lap and nurtures all
> their offspring, mingled great with her great body.

A much earlier example of the generative role of water from the sky is attested, as we saw in the Introduction, in *Rig Veda* 5.58.7, where the Maruts (rain-gods)

> lay their strength in (the earth) as a husband the embryo. They have harnessed the winds like horses to the poles, the sons of Rudra have made their sweat into rain.[9]

Here too the anthropomorphic model of husband and wife is applied to the cosmological level (but not reapplied to the human level): the vigorous Maruts with their rain-clouds fertilize a relatively passive Earth.

Words of the **wers-* family are attested in the Vedic as well as the Greek contexts. In the fragments from Aeschylus and Euripides, as we have seen, the name of the Sky-father is *Ouranós* (**wors-anos*), and in the Vedic verse mois-

----

8 These passages, as well as the fragments of Aeschylus and Euripides, are discussed in a forthcoming article by J. C. Herington on the marriage of sky and earth. The same theme is expressed in similar terms in Statius, *Silvae* 1.2.185-186 and the *Pervigilium Veneris* 60-61. See Motte 1973.220.

9 Translation based on Geldner 1951, part 2.66. See again Onians 1951.191-193, for the essential identity of strength, seed, and sweat.

ture from the Maruts is *varṣám* 'rain', another **wers-* derivative.[10] In fact, of the nine instances where *varṣám* and its compounds are attested in the *Rig Veda*, six are associated either with the Maruts or with Parjanya, another male divinity of the sky.

In Greek traditional poetry, however, apart from passages involving the marriage of Earth and Sky, 'rain' (*ómbros, húetos,* and the verb *húo*) rarely appears as fertilizing moisture. The only examples in Homeric epic of such a connotation are in *Od.* 13.245, where unfailing *ómbros* and *eérsē* account for the flourishing grain and vines of Ithaca, and the similar line *Od.* 9.111=358, in which the *ómbros* of Zeus increases the growth of crops and vines among the Cyclopes. In the agriculturally-oriented *Works and Days* (where *eérsē* is never attested) 'rain' may likewise connote welcome and necessary moisture (e.g. 415, 451, 488-492).

On the other hand, always in the *Iliad*, and often in other early poetry, 'rain' (*ómbros* or *húetos*) has negative connotations. Whether or not overtly associated with Zeus, these words imply destructive storms (*Il.* 10.6, 11.493, 12.286, 13.139) or at least unpleasant weather (*Od.* 4.566, 6.43).[11] 'Rain'

----

10 Unlike *ómbros* and its cognate *imber*, the words for 'rain' in the Greek and Latin passages, which derive from the root **embh-*. This etymological difference is not surprising; Chantraine 1968-77 *s.v.* ὕετος comments on the diversity of words for 'rain' in the Indo-European languages.

11 Such perhaps is the implication of *ómbros* in *Od.*5.480=*Od.*19.442, in passages describing thickets too dense for sun or rain to penetrate.

threatens or destroys trees (*Il.* 12.133), animals (*Il.* 3.4, *Od.* 6.13), and works of men (*Il.* 5.91, 12.250). In several contexts *ómbros* causes wood or bones to rot (*Il.* 23.328, *Od.* 1.161, *WD* 626). The rainy season is dangerous for sailing (*WD* 674-676). Finally, in *Il.* 16.385 Zeus pours down *labrótaton húdōr* 'most furious water' to punish evil-doers. When rain is mentioned apart from generative contexts, then, its associations are usually negative, unlike those of *eérsē* or *drósos*.

Apart from theogonic or cosmological myths, Zeus rather than Ouranos is the Greek god associated with rain, as seen in many of the preceding examples from epic (and also in the common formulation *Zeùs húei* 'Zeus rains' (e.g. in *Il.* 12.25, *Od.* 14.457, *WD* 488). As he sends *ómbros* so also he sends *eérsē*: In *Il.* 11.52-55 'dew-drops dripping with blood' (ἐέρσας αἵματι μυδαλέας) are sent by Zeus as a sign that he will send many heroes to Hades in the ensuing battle. This image is expanded to a thunderbolt accompanied by 'bloody drops' (ψιάδας αἱματοέσσας) in the Hesiodic *Shield of Heracles*, as a prelude to the battle between Heracles and Cycnus (*Sh. Her.* 383-385). Again, when Zeus sheds 'bloody drops' in sorrow for the impending death of his son Sarpedon (*Il.* 16.459-461), the reference is no doubt to rain or dew as the tears of Zeus.[12] The paradoxical nature of 'bloody dew' will be discussed below;[13] at

----

12 Cf. the ῥανίς 'drop' felt by Dicaeopolis in Aristoph. *Ach.* 171 and interpreted as a 'sign from Zeus' (διοσημία).

13 Pp. 74-79.

this point our concern is merely to see in what ways Zeus, who is after all the Greek reflex of the Indo-European Sky-father *Dyēus*, is associated with celestial moisture.

Zeus, whose generative powers are of course notorious in myths, is occasionally connected with the fecundating aspect of heavenly moisture, including *eérsē* and its derivatives. One passage in the *Dionysiaca* of Nonnus, an archaizing epic dated to the fourth or fifth century A.D., overtly represents the fecundating seed of Zeus as *eérsē:* Semele, pregnant with Dionysus, sees in a dream a garden tree (representing herself) struck down by fire. Its fruit, however, "gleaming with the fruitful dew-drops (ἀεξιφύτοισιν ἐέρσαις) of Cronus' son", is rescued by a bird and carried to Zeus to be sewn into his thigh (*Dion.* 7.146ff.). Clearly the 'dew-drops' embody Zeus' male fertilizing powers.

The apparent meaning of *eérsē* in this late passage has a subtle Homeric parallel in another context pertaining to Zeus' sexual activity, the famous Seduction of Zeus by Hera in *Iliad* 14. Beguiled by his wife's beauty when she arrives at his outpost on Mt. Ida, Zeus proposes that they lie together immediately. Hera modestly objects on the grounds that they might be seen by other gods. But Zeus finds a solution: he will cover them with a golden cloud that not even the eyes of Helius can penetrate (*Il.* 14.346-351):

Ἦ ῥα, καὶ ἀγκὰς ἔμαρπτε Κρόνου παῖς ἣν παράκοιτιν·
τοῖσι δ' ὑπὸ χθὼν δῖα φύεν νεοθηλέα ποίην,
λωτόν θ' ἑρσήεντα ἰδὲ κρόκον ἠδ' ὑάκινθον
πυκνὸν καὶ μαλακόν, ὃς ἀπὸ χθονὸς ὑψόσ' ἔεργε.
τῶι ἔνι λεξάσθην, ἐπὶ δὲ νεφέλην ἕσσαντο
καλὴν χρυσείην· στιλπναὶ δ' ἀπέπιπτον ἔερσαι.

He spoke, and Cronus' son took his wife in his arms.
Under them the divine ground grew fresh new grass,
dewy (*hersēenta*) clover and crocus and hyacinth
thick and soft, which held them up from the ground.
On this they both lay down, and drew around them a cloud
fair and golden, and glistening dew-drops fell down.

Although this whole passage has more to do with erotics than with procreation, the "dewy clover" under the divine pair, and especially the "glistening dew-drops" falling on them from a golden cloud, nevertheless suggest the fecundity of the Sky-father. It is as if the cosmological pattern has been displaced by a Homeric tale of Olympian intrigues, leaving its traces not only in the 'dew' of the cloud, but in earth's sudden lush growth of grass and flowers in response to the love-making of Zeus and Hera.

The moist golden cloud in the Seduction of Zeus has analogues in other descriptions of Zeus' love affairs, notably, as Ninck has observed,[14] in the golden rain by which Danae conceives Perseus (e.g. Pindar, *P.* 12.17, *autórutos chrusós* 'gold flowing of itself'.) A striking parallel to the *Iliad* passage is found in Pindar, *Paean* 6.137-140, where a golden mist covers the island of Aegina as Zeus carries the nymph Aegina there to be his bride:

----

14 Ninck 1921.26.

> ... τότε χρύσεαι ἀ-
> έρος ἔκρυψαν κόμαι
> ἐπιχώριον κατάσκιον νῶτον ὑμέτερον
> ἵνα λεχέων ἐπ' ἀμβρότων ...
>
> then golden wisps of mist (*aēr*) covered your shadowy ridge in that place, so that on the immortal bed . . .

In all these passages the subtle but fecund moisture of Zeus ('dew' or 'mist') is present as a *sign* of his generative powers rather than overtly as their vehicle. Such "symbolic" use of dew differentiates the mythical contexts involving Zeus and a particular consort from the cosmological contexts describing the union of Ouranos and Gaia or the like, where 'rain' is actually fecundating liquid. In several poems, however, *eérsē* and other words with similar meanings are used to designate semen--although never in connection with Zeus. A passage in Nonnus (*Dion.* 41.63-64), for example, recalls the remarkable birth of Erechtheus:[15]

> ... ὃν τέκε γαίης
> αὔλακι νυμφεύσας γαμίην Ἥφαιστος ἐέρσην.
>
> ... whom Hephaestus begot
> wedding his marital dew to a clod of earth.

A similar image is attested in a poem attributed to Palladas (sixth century A.D.?) in the *Greek Anthology* (*Anth. Pal.* 10.45) where semen is designated by

----

[15] The familiar birth story usually refers to Erichthonius, as in Apollodorus 3.14.6. Nonnus, however, consistently uses the name Erechtheus, following Homeric usage. See *Il.*2.547, where a scholion says that Erechtheus was also called Erichthonius. For the similarities and differences between the two names, see Burkert 1972.176 and 211, and Loraux 1981.28-29.

rhanîs 'drop', a word which elsewhere is used naturally enough of water from the sky (e.g. Eur. *Andr*. 227, Aristoph. *Ach.* 171). Again, a late Latin allegorical erotic poem equates semen with 'dew' (*Anth. Lat.* 1.2.179-180):

> . . . trepidante cursu venae et anima fessula
> eiaculant tepidum rorem niveis laticibus.
>
> in trembling course the veins and poor tired spirit
> cast out warm dew (*ros*) in snowy liquid.[16]

These late metaphorical attestations of 'dew' as 'semen' recall the semantics of **wers-*, whose primary meaning is (fecundating) liquid from the sky, but which often connotes male reproductive powers as well.[17] This relationship may be explained in part by the visual resemblance of dew to semen,[18] but the cosmological model, the generative waters of the Sky, may have further contributed to the metaphorical equation of the two liquids. This is especially evident in the Nonnus passage, where Hephaestus' 'dew' is 'wedded' to a piece of earth in an image recalling the marriage of Ouranos and Gaia.

In what has been interpreted as a related semantic development, young

----

16 See Clarke 1974 esp. 71-72. Clarke theorizes that Greek *érōs* is cognate with Latin *ros* and other words relating to liquid, and that it originally meant "mist or dew as primal generative moisture." While I am not convinced by this etymology, Clarke's article nevertheless makes a strong case for the contextual associations of dew and generativity.

17 The connection between 'dew' and 'fecundating liquid' is not limited to Indo-European languages; in Chinese folklore, for example, drinking dew may lead to conception. See Thompson 1932-36 *s.v.* 'Dew' T 512.7.

18 As suggested in Clarke 1974.71.

animals are sometimes called 'dew-drops' in Greek. Such is the usual interpretation of *hérsai* in the *Odyssey*, in the passage where Polyphemus divides his flock for the night (*Od.* 9.220-222):

> ...διακεκριμέναι δὲ ἕκασται
> ἔρχατο, χωρὶς μὲν πρόγονοι, χωρὶς δὲ μέτασσαι,
> χωρὶς δ' αὖθ' ἕρσαι...
>
> ... Each kind was separated and
> penned apart, here the forebears(?), here the middlers(?),
> and here the dew-drops(?) ...

Further examples of 'dew' or 'drops' meaning 'young animals' are found in Aeschylus, *Agam.* 141, where lion cubs are called *drósoi;* in Nonnus, *Dion.* 3.389, where *eérsai* is used of young lions again; and apparently in Sophocles fr. 725N, where the phrase *psachaloûchoi mētéres* 'mothers with droplets' probably refers to animals with their young. Nor is this metaphor confined to poetry: Aelian tells us that some people, among them the Thessalians, refer to young birds, snakes, and crocodiles as *psákaloi* 'drops, spots' (*N.A.* 7.47).[19]

A number of scholars[20] have argued that, in effect, this metaphorical use of a variety of words meaning 'dew' or 'drop' results from a misunderstanding of *hérsai* (or *érsai?*) in the *Odyssey* passage, where it is actually a derivative of **ers-* and means 'male animals' (cf. Latin *verres* 'boar'). Whatever the deriva-

----

19 See also LSJ *s.v.* ψάκαλον.

20 E.g. Giles 1889; Denniston and Page 1957.81 (commentary on Aesch. *Agam.* 140-145).

tion of the word for Polyphemus' lambs in the *Odyssey*, however, in the other contexts where a word meaning 'dew' or 'drop' signifies 'young animals' there is no hint that the animals in question are males. On the contrary, a gloss in Hesychius[21] defines the word *stagónes* (related to the verb *stázō* 'drop, shed') as either 'drops (of liquid)' or 'daughters', indicating that in this case at least 'drops' can specify female rather than male offspring.

Beekes provides another kind of argument against identifying *hérsai* in the *Odyssey* passage as a **wers-* derivative, declaring that in *Od.* 9.222 the word cannot have begun with digamma (reflex of IE /w/), presumably because it follows an elision (αὖθ' ἕρσαι). Beekes concludes that *hérsē* is not a figurative use of *eérsē*, but a different word, probably designating an age category of lambs.[22] It should be noted, however, that in *Il.* 14.348 (the Seduction of Zeus), *hersḗeis* is also attested following an elision (λωτόν θ' ἑρσήεντα), yet Beekes does not question the relationship of that word to *eérsē*.

The fact remains that a number of words for 'dew' or 'drop' are used of young animals in Greek. Although Fraenkel found that this image was "peculiar to the Greek point of view,"[23] a similar notion is attested in other

----

21 σταγόνες· ῥανίδες, θυγατέρες.

22 Beekes 1969.64.

23 Fraenkel 1950 vol. 2.83, in his commentary on *Agam.* 141, following Bechtel.

languages as well, e.g. Hungarian *csöp* 'drip, drop', also used of baby animals.[24] The best explanation for the metaphor is still that of Eustathius in his commentary on *Od.* 9.222: *hérsai* means "the youngest, dewy because of their tenderness (*tò hapalón*)". Many attestations of *eérsē*, *drósos* and related words in Greek poetry illustrate the range of meanings 'moist, tender, lively, appealing', from Hecabe's description of the corpse of Hector as *hersḗeis* to the erotic epigrams of Paulus Silentarius praising the lips of his beloved as *droserós* (*Anth. Pal.* 5.244 and 270). The same qualities may have generated a series of Greek names based on words for 'dew' or 'drop', including *Drosîs*, *Psekás*, *Psîax*, *Stagónion*, *Stagṓn*, and *Rhanîs*.[25]

In one context, the dew metaphor is clearly applied to a (mythical) human child. Callimachus, *Hec.* fr. 260.19, apparently refers to Erichthonius as the *drósos* of Hephaestus. With the marked significance of male generative fluid in this myth (as noted above, the seed of Hephaestus is actually called *eérsē* in Nonnus, *Dion.* 41.64, albeit a much later text), it is plausible that Callimachus is here suggesting a correlation between the 'dew' of the male parent and the infant 'dew-drop', as if the generativity of the father were transformed into the life of the offspring. If so, the myth of Erichthonius would provide a striking, perhaps playful example of convergence between these two traditional develop-

----

24 For this information I am grateful to my colleague Blaise Nagy.

25 Bechtel 1917.598-599.

ments of words for 'dew', so that the primary significance of fertile moisture from the sky is reflected both in the fecundity of Hephaestus and the tender vitality of his earth-born son.[26]

One important aspect of the Greek image of generative 'dew' is the emphasis it puts on the male role in conception, whether on the cosmological or the mythical level. Even in the fragments quoted above from Aeschylus and Euripides, where the presence of Aphrodite suggests that both Ouranos and Gaia mutually desire their "marriage", the role of Gaia is merely that of an eager recipient of her husband's fertilizing moisture. With this view we could compare the contrasting role of a self-sufficient Gaia, populating the empty world by parthenogenesis, as in *Theogony* 116ff.[27] Even in the passage quoted from the *Georgics* above, the "mixing" of Aether and Earth in their union may suggest that both parents contribute materially to the formation of their offspring.

Recent studies by Zeitlin[28] and Loraux[29] show that the question of the

----

26 This question will be discussed further in Ch. 5 below.

27 See Arthur 1982 for an eloquent description of the gradual subservience of Gaia to patriarchal goals in the *Theogony*. Aristotle argues against the possibility of female self-sufficiency, at least in terms of procreation, on the grounds that otherwise the male would be useless, and nature makes nothing without some purpose: *Gen. An.* II.5.741b. See further Loraux 1981.90-91.

28 Zeitlin 1978.

29 Loraux 1981.

relative importance of male and female in reproduction was a particularly significant issue in fifth-century Athens. That men might produce children without women is voiced as a desirable ideal by male characters in two Euripidean tragedies: Jason in *Medea* 573-575 and Hippolytus in *Hipp.* 616-619. This ideal is virtually claimed as a reality by Apollo in Aeschylus, *Eum.* 658-661, in his defence of the matricide Orestes:

> οὐκ ἔστι μήτηρ ἡ κεκλημένη τέκνου
> τοκεύς, τροφὸς δὲ κύματος νεοσπόρου·
> τίκτει δ' ὁ θρώισκων, ἡ δ' ἅπερ ξένωι ξένη
> ἔσωσεν ἔρνος, οἷσι μὴ βλάψηι θεός.

> The one called mother of a child is not
> a parent but a nurse of the new-sown fetus;
> the active male begets, but she, like friend for friend
> preserves the seedling, unless a god should harm it.

The dramatic location of the trial scene is Athens, as is of course the site of the tragedy's performance. Apollo's argument--ultimately a successful one--is addressed to an Athenian jury. The living proof of his assertions is the city's own goddess, Athena, whom "a father created without a mother...not nurtured in the darkness of a womb" (*Eum.* 663-665). In view of the emphatically Athenian nature of Orestes' acquittal,[30] we should note that earlier in this tragedy Apollo's priestess refers to the Athenians as "Hephaestids" (*Eum.* 13), a rarely

----

30 Without neglecting the great and fruitful compromise which Athena effects with the Eumenides once Apollo and Orestes have left the scene, I would nevertheless insist that Orestes *is* acquitted--by the deciding vote of Athena herself. It is particularly instructive to compare the resolution of Orestes' dilemma in this tragedy with his fate in other versions of his story, e.g. Eur. *Orestes*.

attested patronymic that must refer to their descent from Erichthonius, here considered as "son of Hephaestus" rather than "earth-born" (as his virtual equivalent Erechtheus is called for example in *Il.* 2.547.)[31] A context where Hephaestus rather than Earth is considered the primal ancestor of the "autochthonous" Athenians is certainly one which places a high premium on the role of the male parent. This same value, or belief, underlies the all-important generative role of a Sky-Father in cosmology. Although in the beginning Earth may have produced offspring by herself, the world as it is now depends on the moist fertility, (whether 'rain', 'mist', or 'dew') of Sky. Moreover, as we have seen, this relationship is used as a paradigm for human relationships, most clearly in the *Danaids* fragment, where Aphrodite presents the fertilizing of Gaia by Ouranos as a paradigm, perhaps to justify Hypermestra's refusal to murder her husband.[32]

The generative nature of moisture from the sky, inherited in **wers-* derivatives such as *Ouranós* and *eérsē* and attested for other words meaning 'dew' or 'rain' as well, is strikingly confirmed on the level of natural history. Although the writings on this subject seem far removed from the realm of early Greek

----

31 Loraux 1981.132 esp. n. 55, argues for the unusual nature of this title of the Athenians, in opposition to the interpretation of Burkert 1977.261, who concludes that the Athenians in general did honor Hephaestus as a paternal ancestor.

32 But Garvie 1969.204-233, after an exhaustive review of the evidence for the structure of the *Danaids*, concludes that the context of Aphrodite's speech is still unknown.

poetry, nevertheless a comparison with them can be productive. Such texts often reflect a cosmology similar to that of traditional poetry, albeit a more "secularized" view that ascribes to natural causes the fertilizing powers which are attributed to immortals in the language of early epic, lyric, and drama.

So it is that authors such as Aristotle, Theophrastus, and Pliny the Elder, see 'dew' (Greek *drósos*, Latin *ros*, etc.) as moisture with a remarkable capacity to generate plant and animal life. In the pseudo-Aristotelian treatise *De Plantis* 2.3.824b, for example, seeds of *botánai* 'wild plants' are said to be produced when vapors ascending from the earth condense and the stars act upon the resulting moist air which *drosízei* 'bedews' an area. In this process the author emphasizes that "the necessary *húlē* 'material' is *húdōr* 'water'."

Certain insects also originate in dew. Aristotle, *Hist. Anim.* 5.19.551a, catalogues various sources for the spontaneous generation of insects; first in the list is *drósos* which has fallen on leaves. The creatures thus produced are said to arise 'from' (*ek*) the dew, just as elsewhere animals are born 'from' other animals of like kind. By contrast, insects are spontaneously generated 'in' (*en*) all the other substances discussed here, such as mud, dung, hair, flesh, or wood. At the risk of overreading the prepositions in this passage, I would suggest that generation 'from dew' was felt to more closely resemble origin from a parent animal, as if the dew itself provided the seed of life, whereas generation 'in' another substance merely specifies the site where life arises.[33] One could further propose a parallel based on the roles of the two sexes in procreation: does the offspring arise 'from' the father but 'in' the mother?

In any case, a detailed description of spontaneous generation in Pliny, *Nat. Hist.* 11.112, illustrates the idea that 'dew' (like semen) contains not only fertilizing moisture but even the generative seed itself. The section is worth quoting in full:

> Many insects however are born in other ways as well, and in the first place from dew (*e rore*). At the beginning of spring this lodges on the leaf of a radish and is condensed by the sun and shrinks to the size of a millet seed (*milio*). Out of this a small maggot (*vermiculus*) develops, and three days later it becomes a caterpillar (*uruca*) which as days are added grows larger; it becomes motionless, with a hard skin, and only moves when touched, being covered with a cobweb growth--at this stage it is called a *chrysalis*. Then it bursts its covering and flies out as a butterfly (*papilio*).[34]

Further testimony to the generative power of 'dew' appears in Aeschylus, *Agam.* 560-562, where Agamemnon's messenger returning to Argos recalls how heavy dews caused the Greeks besieging Troy to become infested with vermin. Moreover, a passage in Pliny suggests that dew can stimulate even inorganic matter to produce. The author cites an old belief that soda-foam (*spuma*

----

33 See Loraux 1981.100-101 for further discussion of ἐκ in contexts of metamorphosis or filiation, esp. concerning the discussion of τὸ ἐκ τινος εἶναι, Aristotle *Metaph.* 1023a26-1023b11.

34 Translation from Rackham 1967 (Loeb Classical Library).501-503.

nitri) could be generated only when dew (*ros*) fell on soda-beds "teeming" (*praegnantibus*) but not yet "giving birth" (*parientibus*) to the substance (*Nat. Hist.* 31.112). In terms of the zoological language of Pliny's description then, dew actually promotes birth, rather than conception, of the soda-foam--a view which underlies the striking image in Aeschylus, *Agam.* 1390-1392, where moisture from Zeus helps a grain-field to give birth to its crop.[35]

Pliny also provides what must be the most comprehensive statement of the fertilizing powers of dew in the writings of natural historians, with his discussion of the risings of the planet Venus (*Nat. Hist.* 2.38):

> huius natura cuncta generantur in terris; namque in alterutro exortu genitali rore conspergens non terrae modo conceptus inplet, verum animantium quoque omnium stimulat.
>
> Its influence is the cause of the birth of all things upon earth; at both of its risings it scatters a genital dew with which it not only fills the conceptive organs of the earth but also stimulates those of all animals.[36]

This remarkable cosmological theory, reminiscent of the descriptions we have seen in Aeschylus, Euripides, Lucretius and Vergil of the union of Sky and Earth, also recalls in a much broader context the combined action of "stars" and moisture cited in pseudo-Aristotle, *De Plantis* 2.3.824b. Significant too is the fact that Pliny explicitly excludes the possibility of supernatural influence from

----

[35] See below pp.73-74 for discussion of this passage from another point of view.

[36] Text and translation from Rackham 1979.193.

the heavenly bodies (*Nat. Hist.* 2.28-29). The generative influence of Venus and *ros* is purely natural.

Finally, Theophrastus too recognizes the fertilizing influence of 'dew' on plant germination and development. In a work on the growth of plants he advises planting in spring or fall, because at those times the earth is moist (*dî ugros*), the sun warm, and the air *malakòs kaì hersṓdēs*, 'soft and dewy'[37] so as to promote the nurture (*ektrophḗ*) and germination (*eublastîa*) of all things (*C.P.* 3.2.6). The passage from Theophrastus, among others, attributes to 'dew' the power not only to generate life but to enhance and increase it. This subject, the nurturing qualities of moisture from the sky, will be discussed in the next chapter.

----

37 Note the unexpected attestation of a derivative of *éersē* rather than e.g. *drosṓdēs* from *drósos*, in a prose work of the fourth century. This indicates at least that the word was intelligible in this period, although normally *éersē* is restricted to poetry. This is especially noteworthy in view of the very "prosaic" quality of Theophrastus' language, on which see Silk 1974.44.

Chapter Two

NURTURE AND TRANSFORMATION

In contexts dealing with generation and birth, as we have seen, *eérsē*, *drósos* and other words signifying 'dew', 'rain' or 'drop' have much in common with the semantics of the rhyming roots **wers*- 'rain' and **ers*- 'male', since the concept of 'liquid distilled from the sky' is often associated with male generative powers in Greek cosmology and mythology. In contexts that deal with the growth and nurture of life already begun, however, these same words are often linked with female agents, especially with goddesses who are called "daughters of Zeus". Thus the Muses dispense *eérsē* to infant kings (*Theog.* 81-83), *Érsa* "daughter of Zeus and Selene" apparently nurtures plants (Alcman fr. 57P), and a family of victorious athletes is said to water their fatherland with "the fairest *drósos* of the Charites" (Pindar, *I.* 6.63-64). Several times, *eérsē* itself is called *thêlus* 'female'. In this chapter we will examine contexts where 'dew' nurtures or otherwise transforms life, again observing with special interest those passages in which it is associated with various immortals.

Agriculture. In early epic, moisture from the sky occasionally encourages the growth of sown or planted crops. In *Od.* 13.244-245, for example, after Odysseus has returned to an Ithaca he does not recognize, Athena describes to him how productive the island is:

> ἐν μὲν γάρ οἱ σῖτος ἀθέσφατος, ἐν δέ τε οἶνος
> γίγνεται· αἰεὶ δ' ὄμβρος ἔχει τεθαλυῖά τ' ἐέρση.
>
> For on it grows grain beyond telling, and wine too,
> and rain keeps it always and flourishing dew.

Only one other context in Homeric epic presents *eérsē* as beneficial to cultivated plants. In *Il.* 23.597-599, *eérsē* is used in a simile to describe how Menelaus, angry at being tricked by Antilochus in the horse race, softens when the winner willingly presents him with the prize mare:

> . . . τοῖο δὲ θυμὸς
> ἰάνθη ὡς εἴ τε περὶ σταχύεσσιν ἐέρση
> ληΐου ἀλδήσκοντος, ὅτε φρίσσουσιν ἄρουραι.
>
> . . .and his temper
> was warmed, as if dew (were warmed) around the ears of grain
> on the growing crop, when the ploughlands bristle.

It is difficult to establish the precise relationship between tenor (Menelaus' temper) and vehicle (dew on the ears of grain) in this simile: how, in particular, does the verb *iaînō* 'warm, cheer' apply to each clause? G.E.R. Lloyd offers a plausible interpretation:

> The sense of ἰαίνω applied to Menelaus' θυμός at 598. . .is 'cheer', but the literal meaning of this verb. . .is 'warm' or 'melt', and this, rather than the derivative 'cheer', is the sense which predominates when the verb is supplied with ἐέρση in the second clause.[1]

If Lloyd's interpretation of *iaînō* is correct, then *eérsē* is here correlated with the notion of 'warmth'. It is this combination of heat and moisture which is implicitly beneficial, or welcome, to the ears of grain in the "bristling" fields.

----

[1] Lloyd 1966.188.

We shall see later that the combination of dew and cold has quite different results.

The idea that 'dew' (as opposed to 'rain') is helpful to sown crops is restricted in early Greek poetry to these contexts, although a similar notion underlies the general statement of Theophrastus that the best time to plant is when the air is "soft and dewy" (*C.P.* 3.2.6).[2] In general, early Greek works say little about the influence of dew on agriculture; for example, *éersē* (like *drósos*) is not once attested in the *Works and Days.* In the agricultural writings of the Roman or Byzantine periods, more attention is given to the potentially harmful effects of dew. Nevertheless, in a few special cases it is considered to be helpful: e.g. for grafted plants needing especially gentle moisture,[3] or for certain kinds of grapevines.[4] The *Geoponica*, a Byzantine work of perhaps the ninth century which was evidently much influenced by Roman agricultural writings,[5] recommends dew (*drósos*) for several purposes: to cure dried figs, cover them in wine lees and soak them in dew before storage (10.54.3); to ren-

----

[2] See above, pp. 11-16, for the influence of rain on crops; p. 30 for a brief discussion of the Theophrastus passage.

[3] Pliny, *Nat. Hist.* 17.25: "inoculatio (i.e. patch-budding) rores amat lenes." On the various kinds of grafts known in ancient horticulture, see White 1970.257.

[4] Columella 3.1.7.

[5] White 1970.15 and 32, analyzes this "disorderly Byzantine compilation" and its sources.

der palm fronds suitable for weaving, expose them alternately to sun and dew (10.6.2). (Here again, dew is considered beneficial when combined with warmth, in this case the heat of the sun.) Both of these processes involve, however, not the nurture of growing plants but the transformation of vegetal materials into useful processed products.

Apart from these limited uses, the prose writers on agriculture generally tend to view dew as harmful moisture. The Roman agriculturalist Columella (first century A.D.) provides the most salient examples of this negative view in his long treatise *De Re Rustica*. He warns his readers about the dangers of "frozen dews" (*concreti rores*), which have dire effects on the health of plants, animals[6] and human beings (1.5.8)--an example of the deleterious effect of dew combined with cold. But for Columella dew can be harmful in any temperature; he cites its dangers in particular to vetch (2.10.29), beans (2.10.10) and grapevines (4.19.2). Pliny the Elder too, in his *Natural History* written at about the same time as Columella's handbook, recognizes *ros* as a threat to certain kinds of vines (14.32) and wheat (18.91), and warns that excessive dew causes 'blight' or 'mould' (*scabies*) in plants (31.33). The harmful effects of dew are explained in detail for fig trees: mild dewfalls (*rores lentes*) cause *scabies*, heavier dewfalls cause the green figs to fall off the tree, and excessive rains (*imbres*) weaken the tree by causing wet roots (17.225). On the farms of central Italy in

----

6 Similarly, *Geoponica* 18.2.7 warns against taking sheep out to pasture before the dew has dried in cold weather.

the first century, then, heavy dewfall seems to have presented real dangers to crops. Moreover, the negative aspects of dew may have been magnified in view of outbreaks of malaria, which apparently threatened parts of Italy at that time. Evidence for this comes from Columella himself, in his warning to avoid marshy lands and standing waters, for these places give rise to creatures too small to see, which enter the body and cause serious disease (1.5.6).[7] Apparently, excessive moisture of any sort was perceived as dangerous by Columella, and his general attitude toward dew should be understood in such a context.

It is clear then that practical writers on Roman agriculture were at best ambivalent about the effects of dew on their crops and animals, and even on human health. A glance at Roman poetry, however, shows that 'dew' on plants or fields connotes pleasant moisture in Vergil and the other Latin poets who would have been most familiar to Pliny and Columella. (Columella, for example, quotes Vergil over fifty times, and actually modelled the tenth book of *De Re Rustica*, written in hexameters, on the *Georgics*.) A cursory review of contexts where *ros* is mentioned in Latin poetry suggests that its semantic range is in fact very similar to that of *eérsē* and *drósos* in Greek poetry,[8] where 'dew' pri-

----

[7] In addition, Pliny the Younger calls the Etruscan coast "gravis et pestilens" (*Ep.* 5.6.1), a description which may refer to the incidence of malaria in that area. See White 1970.69, 410, and 417 for discussion of the evidence for malaria in the works of Columella and Pliny.

[8] For passages where *ros* and its derivatives connote pleasant moisture, see e.g. Vergil, *E.*8.15 (= *G.*3.326): "ros...pecori gratissimus"; *E.*8.37: "roscida mala," *G.*3.337: "roscida luna." Horace, *Odes* 3.4.61: "rore puro Castaliae."

marily evokes the notions of fertility and welcome moisture, even when it is attested in agricultural contexts such as *Il.* 23.597-599 and *Od.* 13.244-245. The distinctions between the usual connotation of 'dew' in Greek poetry (together with the similar perspective in Latin poetry) and the attitude of Columella require some explanation.

In part, this difference in perspectives on 'dew' may be explained by climatic differences, on the grounds that Greece is generally drier than Italy (at least the parts of Italy for which the works of Columella were intended), and that consequently dew would be perceived as a more welcome substance in Greek thought than in Roman. Yet, as can be seen by contrasting the semantics of *ros* in Columella with Latin poetry of roughly the same period, difference in climate alone is not adequate to explain the varying perceptions of 'dew'. What other considerations underlie the differences between genres as far as *eérsē*, *drósos* and *ros* are concerned?

In Greek epic and tragedy, as we have seen in the last chapter, *eérsē* and *drósos* are linked with a divine cosmology, based on the concept of a Sky-Father who sends moisture (rain, dew, mist) from the heavens to fertilize Mother Earth. Such a cosmology has ramifications in agricultural as well as religious thought, of course, and in fact the two can be closely related. In an important study of the archaic Greek attitude toward labor, J.-P. Vernant concludes that

----

Propertius 2.30.26: "rorida antra."

human participants in agriculture are felt to take part in a transcendent order, to which both natural and divine elements also contribute.[9] In traditional Greek poetry, 'dew' appears to link the natural with the divine in agriculture and other realms as well. Like rain, it comes from Zeus, whether for productive or destructive purposes; yet it differs from rain in being a gentle distillation of liquid on the earth, reflecting the pervasive generative powers of Zeus. Not only in the patterns of thought and language inherent in their craft, but also in the experience of their society, composers of archaic Greek poetry are part of a culture that finds the "divine" world almost inseparable from the "natural". In this system the traditional significance--one could even say the Indo-European significance--of fecundating moisture from the sky, is still perceived.

Obviously the perspective of Columella differs a great deal from the archaic view of agriculture proposed by Vernant. The Roman writer and landowner considers agriculture as the work of human industry and knowledge, subject to natural forces but hardly to divine powers. We know that his economic perspective is markedly different from that of the small farmer of archaic Greece,[10] and perhaps in his pragmatism and secularism he is equally far

----

[9] Vernant 1971 vol. 2.16-25: "Travail et nature dans la Grèce ancienne."

[10] In spite of his insistence that the owner be involved with his own land (e.g. 1.7), Columella and those to whom his work is directed were owners of large estates worked by tenant farmers (*liberi coloni*) or by slaves. On Columella's four estates in central Italy (3.3.3 and 3.9.2), see White 1970.26. In addition, see White's succinct discussion of the landowning audience to whom *De Re Rustica* is addressed: pp.26, 402-404. Such

removed from the Zeus-oriented world-view of Hesiodic or Homeric poetry and the society it reflects. In sum, the writings of Columella tend to establish a pragmatic contrast with the perspectives of archaic Greek poetry on dew in agricultural contexts.

Uncultivated Plants. Far more often than with agriculture, 'dew' is connected in archaic poetry with the nurture of plants not cultivated for human food, specifically flowers and pasturage. In *Il.* 14.348-351 (the Seduction of Zeus), we recall, Earth produces for Zeus and Hera a bed of soft grasses and flowers, including "dewy clover" (*lōtós*).[11] In examining this passage from the point of view of male generativity,[12] I suggested that the 'dewdrops' (*éersai*) that fall from the cloud surrounding the divine couple reflect the fecundity of Zeus as Sky-Father, although that role is only alluded to somewhat ironically in this context. Now, focussing on the 'dewy' growth produced by Gaia, we see a slightly different aspect of the same theme: Earth, herself nurtured by fertilizing moisture from heaven, now grows plants abounding in that same moisture.[13]

"investment farmers" were far removed economically from the subsistence-level farming of archaic Greece.

11 This phrase may be reflected in Sappho fr. 95.12LP, where the banks(?) of the Acheron are described as λωτίνοις δροσόεντας 'full of *lōtós* and *drósos*'. This passage, which apparently does not involve pasturage, will be discussed below, p.60.

12 See above, pp. 17-18.

13 This concept is taken up more explicitly in Lucr. 2.991: when Earth

In this context, the *lōtós* is to be seen as a wildflower like the *krókos* and *huákinthos* produced by Earth, but elsewhere in Homeric epic, the same plant (not qualified as 'dewy') is fodder for horses (*Il.* 2.776, *Od.* 4.603). In another epic context a different plant used as fodder is called *hersḗeis:*[14] while preparing to sacrifice some of his stolen cattle, the infant Hermes feeds them and drives them into a barn, as they are "chewing on *lōtós* and dewy galingale" (ἐρσήεντα κύπειρον: *H.Hermes* 107).[15]

'Dewy' pasturage then feeds horses and cattle; not far removed from this image is the notion that the pastures themselves are fed by 'dew' from the sky. This idea gains support from a bit of lore preserved in Aelian. A special kind of sweet dew, hyperbolically called *mḗli hugrón* 'liquid honey', falls from the sky in spring onto certain pastures in India. Cattle and sheep are led to places "where the sweet dew (*drósos*) lies fallen even more abundantly". As they feed there, their milk becomes so sweet that, according to Aelian, it can be drunk without the addition of honey (*N.A.* 15.7). Here, on a purely natural level, we find a clear connection between the dew which falls on the pasture and the milk of

----

receives the drops of moisture from Father Aether, she brings forth all forms of life.

14 So too in Latin poetry, e.g. Vergil, *E.*8.15 and the nearly identical *G.*3.326: "ros in tenera pecori gratissimus herba."

15 Again in this passage, dew is apparently involved in a cycle of mutual nurture between heaven and earth. Dew, which falls from the sky (or specifically from Zeus, as in *Il.* 11.53-54, etc.) fertilizes the pasture; cattle eat the dewy grasses; men kill the cattle and render the smoke of sacrifice back up to the heavens.

animals that feed on the dewy grasses. In several respects, this process is much like the one described on the cosmological level as the marriage of Sky and Earth, which generates and nurtures all things that grow, but the "pasturage" contexts place more emphasis on the nurturing than the generative capacity of dew. These two aspects, however, should not be seen as rigidly opposed to each other, but rather as different parts of a continuum reflecting the uses of *eérsē*, *drósos* and Latin *ros* in the semantic range of 'fertility'.

As with grasses, so too with flowers: dew frequently nurtures these uncultivated plants as well. In addition to the 'dewy *lōtós*' produced by earth in *Il.* 14.348, Sappho fr. 95.12-13LP refers to both *lōtós* and *drósos* on the banks of the Acheron. Another fragment of Sappho, better preserved than fr. 95, compares the beauty of an absent woman to a moonlit night when "the fair dew (*éersa*) is shed, and roses bloom and tender chervil and flowery honey-clover" (fr. 96.12-14LP).[16] If the editors' restoration of *eérsē* is correct, dew is again mentioned in connection with flowers in Hesiod fr. 26.20-21MW, which describes the daughters of Porthaon going out early to look for flowers as adornment for their hair.[17] Dew and flowers appear together in a metaphor in Pindar, *N.* 7.79, where the Muse "plucks a delicate flower (*leîrion ánthemon*) from the sea-dew (*pontîa eérsa*)." Whatever this "flower" may be,[18] the fact

----

16 This fragment will be discussed more fully in a later section, pp.58-60.

17 Again, this passage is treated in more detail later: pp.57-58.

that it is taken from the 'sea-*dew*' rather than the 'sea' may reflect how closely dew and flowers are linked in early Greek poetry. As might be expected, this collocation recurs in the pastoral poetry of a later age as well, although not often; for example, in the Theocritean corpus an anguished shepherd blushes "like a rose with dew" (*érsa*: Theoc. 20.16).

The relatively frequent association of *eérsē* and *drósos* with flowers and grasses, even more than with crops such as grain, suggests that dew is associated more closely with nature than with culture, even agriculture. Rain and dew come to earth from the sky (or from Zeus) without human intervention or control. So too, plants fed by dew or rain may grow even without cultivation, simply at the dispensation of Zeus. In a pre-agricultural society, such as that of the Cyclopes, all vegetation is fed by rain or dew alone (*Od.* 9.106-111):

> Κυκλώπων δ' ἐς γαῖαν ὑπερφιάλων ἀθεμίστων
> ἱκόμεθ', οἵ ῥα θεοῖσι πεποιθότες ἀθανάτοισιν
> οὔτε φυτεύουσιν χερσὶ φυτὸν οὔτ' ἀρόωσιν
> ἀλλὰ τά γ' ἄσπαρτα καὶ ἀνήροτα πάντα φύονται
> πυροὶ καὶ κριθαὶ ἠδ' ἄμπελοι, αἵ τε φέρουσιν
> οἶνον ἐριστάφυλον, καί σφιν Διὸς ὄμβρος ἀέξει.
>
> We came to the land of the Cyclopes, overbearing, lawless folk,
> who, trusting in the immortal gods,
> plant no plants with their hands nor do they plough,
> but unsown and unploughed everything grows for them:
> wheat and barley and grapevines which bear
> wine in rich clusters, and Zeus' rain (*ómbros*) makes them grow.

----

18 Although the "flower" is usually understood as a reference to coral, see Petegorsky 1982.141-174, for a detailed argument that the "flower" is really sea-purple dye. See also below, pp.93-94, for further discussion of the Pindaric context.

Such a description recalls in some respects the happy life of the Golden Race of men, who enjoyed all the fruits of the earth without labor, but who, unlike the Cyclopes, were dear to the gods (*WD* 111-120), before the gods hid 'life' (*bíos*) in the earth and mortals had to work for survival (*WD* 42 and *passim*).

A relationship among the gods, rain or dew, and the growth of plants may also lie behind the more limited fertility of Ithaca, where wheat and vines are made to flourish by constant *ómbros* and *eérsē* (*Od.* 13.244-245). The fecundity and prosperity of Odysseus' narrow island is an important theme in the *Odyssey*,[19] although in contrast to other places described in the epic (the dwellings of Calypso and Circe, the palace and gardens of Alcinous, the plain of Elysium where Menelaus will go instead of to Hades), its fertility is of a purely natural order. In Ithaca then we find a good example of how natural fertility which comes from the gods combines with human labor[20] in a world that depends on agriculture and animal husbandry. In this respect the land of Odysseus is opposed to that of the Cyclopes, who simply "trust in the gods" and whose "crops" are increased only by the moisture that comes from Zeus, as are uncultivated pastures and flowers.

----

[19] See especially *Od.* 14.96-104, 20.211-212 for the abundance of herds and flocks, and 19.107-114 where Odysseus compares Penelope to a just king whose lands and people prosper.

[20] See e.g. the comments of Eumaeus, *Od.* 14.63-66, emphasizing the labor he expends in caring for Odysseus' swine.

Dew as Food. In the Hesiodic *Shield of Heracles* 393-397, an elaborate periphrasis identifies the season when Heracles and Cycnus met in battle:

> ἦμος δὲ χλοερῶι κυανόπτερος ἠχέτα τέττιξ
> ὄζωι ἐφεζόμενος θέρος ἀνθρώποισιν ἀείδειν
> ἄρχεται, ὧι τε πόσις καὶ βρῶσις θῆλυς ἐέρση,
> καί τε πανημέριός τε καὶ ἠῷος χέει αὐδὴν
> ἴδει ἐν αἰνοτάτωι, ὅτε τε χρόα Σείριος ἄζει.
>
> when the dark-winged resonant cicada (*téttix*),
> perched on a verdant branch, begins to sing summer for men--
> for him is female dew (*thê lus eérsē*) both food and drink--
> all day long and at dawn too he pours out sound
> at the peak of the heat, when Sirius parches the skin.

We note again the collocation of 'dew' and heat, with beneficial results. The epithet *thê lus* 'female' is appropriate here, too; for although dew typically comes from Zeus, and is often closely associated with male generativity, as food and drink for the cicada it fulfills a typically "feminine" role--providing nurture--as suggested even by the derivation of *thê lus* from the root **dhē*- 'suckle'.

The cicada is one of several small animals that were believed to feed only on dew. Others include snails (Varro, *R.R.* 3.14.2)[21] and a kind of chameleon called the *stellio* (Pliny, *Nat. Hist.* 11.91). But as a dew-eater the cicada (Greek *téttix*) certainly takes pride of place. Its dietary habits are discussed at length in Aristotle, *H.A.* 4.7.523b, a passage closely followed by Pliny, *Nat. Hist.* 11.93-94 (although it should be noted that Pliny sounds somewhat skeptical in

----

21 At least snails ordinarily eat dew; but if no dew falls, they live on their own 'juice' ("suo sibi suco vivont, ros si non cadit": Plautus *Capt.* 81).

reporting that the insect eats nothing but dew.)[22] According to these accounts, the cicada has no mouth, but takes in dew with a tongue-like organ on its underside.[23] Further, it leaves no residue or excrement, thanks to the purity and delicacy of its food.[24] Now according to Dicaearchus, the lack of excrement is a characteristic of men of the Golden Age, along with closeness to the gods, freedom from agriculture, etc.[25] In this respect, the cicada through its diet of dew shares some of the ethereal characteristics of the Olympian gods: both refrain from the corrupting food of the earth. A passage in the *Phaedrus*, moreover, implies a popular belief that cicadas eat nothing at all: Socrates relates the story that originally cicadas were mortal men, who became so dedicated to music that they neglected to eat or drink, but sang continually. When they died (an event they did not even notice!) the Muses allowed them to con-

----

22 Cicadas actually feed on the sap which they suck from leaves. See Snodgrass 1967.200 and Egan 1984.175. See Borthwick 1966 for a fascinating compilation of information on the diet of cicadas and other "grasshoppers" in antiquity, a subject treated in numerous later Greek poems, and known in Latin poetry as well, cf. Vergil, *E*. 5.77. For ancient and modern explanations of the mechanism by which the insect "sings", see Bodson 1976.

23 Cicadas have a beak-like mouth with which they puncture leaves and ingest the fluids extracted. The "tongue-like organ" may be a reference to the ovipositor on the underside of the females. See Snodgrass 1967.199-200.

24 This belief is quite fallacious. Like aphids (Snodgrass 1967.155) cicadas excrete large quantities of a sweet liquid called 'honey dew' which covers the leaves they feed on. Egan 1984.176 correctly concludes: "It is surely this phenomenon which accounts for the widespread notion that cicadas feed on dew."

25 Dicaearchus as quoted in Porphyry, *De Abstinentia* IV.2. See further Detienne 1963.143-145.

tinue their musical career as cicadas, eating and drinking nothing from birth until death (Plato, *Phdr.* 259b-c). In addition to explaining the origin and some of the characteristics of the cicada, this tradition particularly emphasizes the importance of the creature's abstinence from normal terrestrial food: it eats either dew (as in the natural historians and often in poetry),[26] or alternatively, nothing at all. So far is dew removed from a usual diet!

An interesting piece of folklore about the Amazons, preserved in Flavius Philostratus, *Heroicus* 19.19, tells us that these warlike women feed their infants with mares' milk and with the *kēria drósou* '(honey)combs of dew' which settle on river reeds. As with the ethereal cicadas, dear to the Muses, so here too diet reflects the nature of the Amazons. The milk of animals is a substitute for mother's milk, perhaps recalling the popular derivation of *Amazōn* from *a-mazós* 'without breast'. The fact that it is mares' milk, however, has special significance, for the Amazons are characterized by strength and ferocity, qualities attributed to mares in Greek lore.[27] The "(honey)combs of *drósos*" are also significant in terms of the Amazons' diet. This 'dew' forms on uncultivated plants; not only does it differ from the usual diet of infants, but it suggests the 'anti-cultural' life of the Amazons themselves, a subject which has generated a

----

26 See Borthwick 1966 for a full discussion. The connections among cicadas, song, and dew will be taken up again, pp. 67 and 81-84.

27 See esp. the man-eating mares of Diomedes, Apollod. 2.5.8, and note that mares were often champion racers (*Il.* 23.409).

great deal of interest lately. In a recent study of the Amazons, for example, duBois describes how their rejection of marriage (although not of sex altogether) is perceived as a rejection of culture in general.[28] Further, with regard to the cultural significance of diet in particular, Detienne's structuralist interpretation of the dietary aspects of Greek marriage rites shows that for women, the concepts of married life and cultivated grains are inseparable.[29] The Amazons, however, participate in neither marriage nor agriculture. To continue their species, they rely on infrequent and haphazard sexual unions; to feed their infants, they appropriate the milk of horses and the 'dew' of wild plants, preparing them for a life independent of cultural institutions.

Eaters of *drósos*, then, are far removed from those whose diet depends on agriculture, but removed in two different directions. On the one hand, with its ethereal food, the cicada resembles the immortals to whom it is so dear; on the other hand, the Amazons' first foods are emblematic of their undomesticated mode of life. In both cases, *drósos*, whether it is believed to fall onto leaves or to settle on river reeds, is an appropriate food for those dissociated from the cultural norm of mortals.

Sweet Dew and Honey. In the dewfall of "liquid honey" that produces such

----

28 DuBois 1982 esp. 36-39. The passage from Philostratus is not cited by duBois.

29 Detienne 1977.116-117.

sweet pasturage in Indian (Aelian, *N.A.* 15.7) and perhaps in the Amazons' (honey)combs of *drósos*, there is a hint of the sweetness often associated with dew in Greek lore. Further examples of this characteristic are found in the agricultural writers and natural historians. The *Geoponica*, for example, states that dew sweetens wild cucumbers and wormwood (2.36.4)--another case where dew improves uncultivated plants. In a curious passage, Pliny maintains that rains (*pluvia*) sweeten salt, but dewfalls (*rores*) make it even sweeter (*suaviorem: Nat. Hist.* 31.85). In this context, then, the waters of rain and dew are differentiated according to the sweetness each imparts.

The most important aspect of the sweetness of dew is its relationship to honey. The belief is widely attested that honey originates in a kind of 'dew' collected by bees. This concept is apparently quite old; West[30] finds it reflected in the *Theogony*, where the Muses pour 'dew' (*eérsē*) on the lips of kings as they are born, so that *épea meîlicha* 'honeyed words' flow forth (81-84). To Hesiod too is attributed the idea that *robora* 'hard-oaks' produce honey (Pliny, *Nat. Hist.* 16.31). Pliny explains this by referring to *rores mellei* 'honey-dews' that fall from the sky and settle on oak leaves. The connection between honey and dew is probably implicit in Seneca *Ep.* 84, and Columella 9.14.20, where honey is said to come from *cerae* 'waxen cells' (reminiscent of the *kēría drósou* on the river reeds among the Amazons) that appear on flowers with the morning dew.

----

30 West 1966.183 (on *Theog.* 83).

In the most memorable ancient passage dealing with the formation of honey, Pliny describes how honey-dew falls from the sky, especially in the period just before dawn when Sirius is blazing so that "at dawn the leaves of trees are found dewy with honey (*melle roscida*)". Unfortunately, this honey-dew is corrupted in its fall and further deteriorates in the stomachs of the bees which collect and transform it. Pliny longs for the original, unadulterated moisture (*Nat. Hist.* 11.30-31):

> utinamque esset purus ac liquidus et suae naturae, qualis defluit primo. nunc vero a tanta cadens altitudine multumque dum venit sordescens et obvio terrae halitu infectus, praeterea e fronde ac pabulis potus et in utriculos congestus apium--ore enim eum vomunt, ad hoc suco florum corruptus et alvi vitiis maceratus, totiensque mutatus, magnam tamen caelestis naturae voluptatem adfert.

> And would that it were pure and liquid and genuine, as it was when it first flowed down. But as it is, falling from so great a height and acquiring a great deal of dirt as it comes and becoming stained with vapor of the earth that it encounters, and moreover having been sipped from foliage and pastures, and collected into the stomachs of bees--for they throw it up out of their mouths, and in addition being tainted by the juice of flowers, and soaked in the corruptions of the belly, and so often transformed, nevertheless it brings with it the great pleasure of its heavenly nature.[31]

Pliny's remarkable description of the the descent of pure, sweet moisture from the sky and its subsequent contamination with earthly impurities shows many similarities to early Greek notions about dew: its sweetness, its untainted nature (reflected for example in Greek lore about the cicada's diet), its associa-

----

31 Text and translation (slightly adapted) from Rackham 1967.450-453.

tion with Sirius and with dawn (as in *Sh.Her.* 395-396), and of course its pure heavenly origin.

Dew and Daughters of Zeus. An ancient parallel, in the language of poetry and sacred cosmology, to the sweet dew described by Pliny on a naturalistic level, may be found in the *Rig Veda*, in several hymns addressed to *Uṣas* 'Dawn', the daughter of *Dyāus* 'Sky'. She brings "sweet things" (*R.V.* 3.61.5), and "drops moisture like milk" (*R.V.* 7.41.7 and 7.80.3). Especially since dewfall is heaviest at dawn, it is likely that the sweetness and moisture bestowed by the Dawn-Goddess on the divine level reflects dew on the natural level. Just as in a context of male fecundity 'dew' may suggest semen, when the context deals with nurture, especially as dispensed by a female figure, 'dew' may be analogous to milk or honey.

In Greek poetry too, daughters of Zeus (cognate with Vedic *Dyāus*) dispense 'dew'. When the Muses distill *eérsē* on the lips of royal infants, they are explicitly called *Diòs koûrai* 'daughters (or girls) of Zeus' (*Theog.* 81). The kings themselves, moreover, are called *diotréphees* 'Zeus-nurtured' (82).[32] In fact, this whole passage could be considered a gloss[33] on the epithet 'Zeus-nurtured'.

----

32 The formula 'Zeus-nurtured kings' is common in Homeric epic as well, occurring in a series of cases and metrical shapes, e.g. διοτρεφέες βασιλῆες in *Il.* 2.445, *Il.* 14.27; διοτρεφέων βασιλήων in *Il.* 2.196, *Il.* 4.338, etc.

33 See Whallon 1969.65 and *passim*, for the concept of a scene which explains or "glosses" an epithet.

The Muses, as daughters of Zeus, feed heavenly liquid to divinely-favored rulers, who are themselves *ek Diós* 'from Zeus' (96). The *eérsē* originates with Zeus, but is dispensed by his daughters as patron goddesses, and its function is not so much generative as transformative for the kings and their speech.[34] Moreover, it is dispensed for human benefit, since the kings in their turn will use sweet words to restore harmony among their people (85-93). Benefits also accrue from female dispensers of 'dew' when the Charites (who are daughters of Zeus in *Theog.* 907-909) preside over the *drósos* of praise and celebration with which a family of athletes 'water' their paternal land (Pindar, *I.* 6.63-64).[35]

Another daughter of Zeus is attested in Alcman fr. 57P, a fragment quoted by Plutarch in *Quaestiones Conviviales* 3.10.3.

> *οἷα Διὸς θυγάτηρ Ἕρσα τράφει*
> *καὶ Σελάνας*
>
> such things as Ersa nurtures, the daughter of Zeus
> and Selene. . .

The context cannot be determined, but it is most likely to involve the nurture of uncultivated plants. 'Dew' is here personified as a daughter of Zeus and the moon or Moon-Goddess Selene, playing upon the naturalistic relationships among sky, moon and dewfall. Still it is unlikely that for Alcman this passage

----

[34] Arthur 1983.111 seems to suggest that in a sublimated sense the Muses may be "impregnating" the king in the manner of bees, but I believe that the sense of "nurturing" or "transformative" dew prevails here.

[35] See below, pp.95 and 122, for analysis of these passages from a different point of view.

is simply a playful "riddle", as Plutarch suggests in his discussion of the lines.[36] Rather, the full meaning of the lyric fragment surely involves *both* the divine and naturalistic levels of meaning: *Érsa* is both the natural product of air and moonlight, and a divine being, the offspring of Zeus but with a nurturing capacity of her own. In accordance with the mediating nature of 'dew', she conveys fertilizing moisture from sky to earth, spanning the distance between the two realms.

----

36 "Riddling (αἰνιττόμενος) that dew (δρόσος) is the daughter of the air and the moon."

Chapter Three

DEVELOPED IMAGES OF DEW

In the two preceding chapters we have studied the most important connotations of *eérsē* and *drósos* as 'dew' in our sense, that is, (drops of) moisture distilled from the air. The generative or nurturing qualities attributed to dew account for its most common range of meanings in early Greek poetry, and recur in prose genres as well--although here they are attested in more strictly naturalistic contexts.

Now we are ready to examine some of the secondary uses of *eérsē* and *drósos*. In a number of passages the primary significance of 'dew' as fertilizing moisture distilled from sky or air is contrasted in some way with the connotation the context demands. Such deviation from the normal, expected significance of *eérsē* or *drósos* will produce various kinds of "heightened" language: irony, paradox, metaphor. As we shall see, it may even reflect a degree of semantic shift[1] in the meaning of the words, especially of *drósos*. We will try to determine what role such deviation plays in its immediate context, and where relevant also in the larger text of which it is a part. We are thus attempting, in limited scope, what G. P. Landow calls "an archaeology of imagination".[2] Lan-

----

1 For a description of "semantic shift" and its causes see Waldron 1967.142-161.

dow advocates a careful study of figures which a society adopts as "codes", by which he means configurations that refer beyond themselves (as paradigm, synecdoche, analogy, metaphor) to convey an important meaning to the members of that society. On the small scale of the single evocative word, *eérsē* or *drósos* presents such a configuration. The purpose of a careful comparative study of such an image is to recreate the way a text including it was perceived by its original audience, according to Landow, so as to permit us "to differentiate more precisely than otherwise possible apparently similar applications" of the encoded figures (in our case, a context where 'dew' is used in a paradigm, metaphor, etc.). Moreover, this "archaeology" properly applied should allow us "to enter the imagination of another age" without falling "prey to essentially unsupported and unsupportable generalizations about the mind of an age or its *Zeitgeist.*"

Before proceeding with this "excavation" of developed dew imagery in early Greek poetry, it would be well to consider why *eérsē* and *drósos* are so often used in paradoxical, metaphorical or otherwise heightened language. I believe that the answer lies in the nature of the "normal" semantic range of words for 'dew', which are associated with themes of great interest--and great ambivalence--for Greek society. The cosmological origin of life, the relationship between male and female in sexual generation, the necessary subjection of human labor and success in agriculture to the divine vagaries of climate and

----

2 Landow 1982.15-16.

weather, the possibility of rain and dew being communications of Zeus' pleasure or displeasure--all such problems belong to the primary, literal semantic range of *eérsē* and *drósos*. In the sections which follow, I have divided "secondary" or "developed" associations of these and related words into several different categories, each of which can be seen to relate rather closely to at least one of the "primary" contexts, developing some of the ambiguity latent in the image. Admittedly, I have not attempted to discuss at length *all* the secondary uses of 'dew' in Greek poetry, but have selected those which seem to be connected to important themes in the texts of which they are a part.

Erotics. In examining the generative contexts of 'dew', especially in the cosmological sense of the union of Sky and Earth, we have already come upon some passages where a subtle, fecund moisture from the sky (dew or mist) reflects the procreative powers of Zeus. To begin with the latest and most obvious example, in Nonnus, *Dion*. 7.146ff. (Semele's dream) *éersai* 'dewdrops' gleaming on the tree represent in allegorical fashion the fertile seed of Zeus in Semele, which has produced the "fruit" Dionysus. Pindar, *Paean* 6.136-140, recalls the rape of Aegina by Zeus, which culminates as "golden wisps of mist (*aḗr*)" cover the island. In this context too the seed of Zeus is generative, for Aegina will bear Aeacus to him, but compared to the passage in Nonnus, there is greater emphasis on the erotic union of god and nymph: it is during their love-making that mist envelops the island. In the earliest passage where dew or mist accompanies Zeus' sexual activities, the union is solely an erotic one. Hera's laborious preparations for her seduction of Zeus are totally successful. The god, completely beguiled and eager to lie with his wife, sets the scene for their love-making: while Earth grows a soft bed of 'dewy' *lōtós* and other flowers, Zeus draws around them a thick golden cloud whose 'dew-drops' (*éersai*) drip down on the couple (*Il.* 14.346-351).[3] Given Zeus' traditional role of Sky-Father; given the importance of the union of Sky and Earth on the cosmological level, as well as the marriage of Zeus and Hera on the level of ritual; given the connotation of fertilizing moisture inherited in *éersē*--it seems clear that this passage suggests the sacred and generative marriage of a divine father and mother.[4] Yet in the *Iliad* we are dealing not with a primordial pair, as in Aphrodite's description of Ouranos and Gaia in the *Danaids*, for example, but with two highly individualized and personally-motivated Homeric Olympians. Instead of being procreative, their union is only erotic; moreover Zeus is led astray by Hera's guile into thinking that the whole thing is his idea. The multiple ironies of this scene are enhanced by the presence of *éersē*, with its rather misleading connotations of the fecundating powers of the Sky-Father.

Descriptions such as this prepare the way for the use of *éersē* or *drósos* in erotic contexts in later Greek poetry. In addition to the divine or cosmological paradigm, of course, 'dew' is perceived as appropriate for such contexts because of an easy shift from notions such as 'moist, fresh'--appropriate to that which is

----

3 Quoted above, p.18.

4 See above pp.3-5 for the connotations of **wers-* and Ch. 1 for other passages where rain or dew are connected with generation of life.

literally bedewed--to 'soft, tender, appealing'--used of things which are metaphorically 'dewy'. In later Greek erotic poetry, images such as 'dewy lips' (*A.P.* 5.269) and 'dewy mouth' (*A.P.* 5.244) are not uncommon. A final factor contributing to the developed use of 'dew' in erotic contexts is, as we shall see, the fact that human sexual secretions were seen to resemble dew.[5]

A passage in Euripides' satyr-play *Cyclops* seems to exploit some of the erotic implications of *drósos*. The chorus of satyrs try to entice the drunken Polyphemus to go back into his cave by telling him that a tender nymph is waiting for him "inside the dewy caverns" (δροσερῶν ἔσωθεν ἄντρων , 515-516). Euripides' audience may have perceived an erotic connotation in *droserós* here, rather than simply a naturalistic allusion to the moist and cool cave. Perhaps coincidentally, a similar image is found in Propertius 2.30, a poem which could be called the Propertian *Vivamus atque amemus*. The poet invites Cynthia to join him in a pastoral setting that includes *rorida antra* 'dewy caves', and his intentions are clearly amatory (2.30.26).

We have previously discussed contexts where the 'rain' of Ouranos or the 'dew' of Zeus or Hephaestus are analagous to semen.[6] In Aristophanes, whose sexual vocabulary is prodigious, *drósos* may connote the sexual secretions of ordinary mortals. In the parabasis of the *Knights*, the chorus singles out a cer-

----

[5] This possibility is exploited by Aristophanes; see below.

[6] See above, pp. 11-14, 17, 19.

tain Ariphrades, charging that he "licks the disgusting *drósos* in whorehouses" (1285). In the great *agón* of the *Clouds,* Just Logic recalls with nostalgia how in the old days Athenian boys never used to oil themselves below the navel, "so that *drósos* and down bloomed on their privates just as on apples" (978). Dover interprets this 'dew' to mean a pre-ejaculatory genital secretion,[7] but Henderson argues more convincingly that it is rather the appealing moisture of athletic sweat.[8]

In several other contexts, literal 'dew' contributes in an evocative way to the atmosphere suggested by other elements in a description; it forms part of a configuration that suggests rather than describes the erotic. Such is the case, I believe, in Hesiod fr. 26.18-21MW,[9] which describes the daughters of Porthaon, virgins who have refused the works of Aphrodite:

> *αἵ ῥα τότ' εἴδει ἀγαλλόμεναι καὶ ἀιδρείῃισιν*
> *ἀμφὶ περὶ κρήνην . . . ἀργυροδίνεω*
> *ἠέριαι στεῖβον . . . ἐέρσην*
> *ἄνθεα μαιόμεναι κεφαλῆις εὐώδεα κόσμον.*

> glorying then in their beauty and innocence,
> around a spring . . . of silver eddies,
> covered in mist they stepped. . .dew,
> seeking flowers, a sweet-smelling adornment for their heads.

Dew and flowers reflect the girls' own fresh beauty but also their latent

----

[7] Dover 1968 on line 978.

[8] Henderson 1975.145 note 194.

[9] See the important discussion of this fragment in Arthur 1983.98-99, which is also the source (slightly modified) of my translation.

fecundity. Especially in such a context, rejection of the works of Aphrodite is vain. In Aeschylus fr. 44N,[10] Aphrodite herself appears to argue for the importance of human marriage on the model of the fruitful union of Sky and Earth. In the Hesiodic passage, however, the cosmological model of marriage is present in the landscape itself, with the dew still shining on the flowers which it has caused to grow.

After such a description the rape that follows (like that of Persephone from another flowery meadow) is almost predictable. Apollo sees the girls, carries one of them away as bride for his son Melaneus, and soon she becomes the mother of Eurytos (fr. 26.22-28MW). The cosmological imperative is followed; the latent fecundity of the virgin amidst flowers is realized.[11]

Sappho, fr. 96.7-14LP, in a simile comparing a beloved but absent woman to the moon, describes another verdant, flowery scene:

> *. . . ὤς ποτ' ἀελίω*
> *δύντος ἀ βροδοδάκτυλος σελάννα*
> *πάντα περρέχοισ' ἄστρα· φάος δ' ἐπί -*
> *σχει θάλασσαν ἐπ' ἀλμύραν*
> *ἴσως καὶ πολυανθέμοις ἀρούραις·*
> *ἀ δ' ἐέρσα κάλα κέχυται τεθά -*
> *λαισι δὲ βρόδα κἄπαλ' ἄν -*
> *θρυσκα καὶ μελίλωτος ἀνθεμώδης.*

. . .as, when the sun goes down, the rosy-fingered moon

----

[10] Quoted above, p.11-12.

[11] See Motte 1973.198-232 for an exhaustive treatment of the erotic significance of meadows and gardens in Greek literature and ritual.

> surpasses all the stars, and directs her light onto salt sea and flower-laden fields alike. Lovely dew pours down, roses bloom and tender chervil and flowery honey-lotus.

The presence of dew (*eérsa*) enhances in several ways the gentle, feminine atmosphere established in the poem.[12] First, in contrast with the barren salt water of the sea (line 10), the fair dew falls on blooming fields,[13] with the implication that the dew contributes to their flourishing. At the same time, the moon to which the absent woman is compared dominates the night sky, suffusing the land and sea with subdued but effective light.[14] Stigers finds in this simile a reflection of the "female world of Sappho's *thiasos* in its otherness", and implies a connection between the masculine world and the sun, which are absent from the mysterious, private world of the poem. The moon here may be seen as a sun *manqué*, and dew as rain *manqué*. We may note even the feminine gender of substantives for 'moon' and 'dew', as compared to the masculines for (absent) 'sun' and 'rain'. Further, in this simile the flower-nourishing *eérsa*, associated as it is with the "rosy-fingered moon", subtly suggests the pervasive effects of the moon-like woman's presence--or perhaps the memory of her in her

----

12 Stigers 1977.93.

13 See McEvilley 1973, esp. 264-266 and 274-275, for a detailed interpretation of the nature imagery in this poem.

14 *Quaestiones Morales* 24, where Plutarch quotes Alcman, fr. 57P ("such as Ersa nurtures, the daughter of Zeus and Selene"), provides an interesting background to the collocation of moon and dew. Plutarch contends that dew (*drósos*) is "weak and feeble rain (*ómbros*), and weak too is the heat of the moon, so it draws moisture from the earth as does the sun, but being unable to carry it up high or to maintain it, releases it."

absence. To call this description "erotic" would be too forceful; rather, the flowers, moon and dew subtly suggests a landscape suitable for lovemaking, its private feminine beauty connected by analogy with the absent woman whose beauty still affects those under the spell of her memory.[15]

'Dew' appears also in Sappho, fr. 95.11-13LP, a badly fragmented poem:

> κατθάνην δ' ἴμερός τις [ἔχει με καὶ
> λωτίνοις δροσόεντας [ὄ-
> χθοις ἴδην Ἀχερ[
>
> A desire to die . . . (and) to see the clovery dewy banks of Acheron. . .

The collocation of *lōtós* and *drósos* on the banks of the Acheron is reminiscent of the 'dewy clover' (*lōtòs herséeis*) produced by Earth in *Il.* 14.348, but it is surprising to find such a verdant description applied to a river of Hades. It seems that in this fragment, in the context of a wish for death, the underworld is visualized paradoxically as a place of lush fertility.[16]

Pure Water, Celestial and Terrestrial. Dew is traditionally connected with Zeus, or sometimes with various daughters of Zeus.[17] As a result of its divine and celestial origin, it is associated with notions of ethereal purity and closeness to

----

15 See now Hague 1984.31, who notes the similarities between Sappho fr. 96 and *Il.* 14.347-351, concluding that "the emphasis on fertility and growth is furthered by the sprinkling of dew."

16 See Boedeker 1979 for a detailed analysis of fr. 95 from this perspective.

17 See above, pp. 31, 49-51.

the gods. We have already seen how the cicada, which eats only dew, is favored by the Muses,[18] and how Pliny, describing the origin of honey from dew, waxes uncharacteristically poetic in describing the pure, untainted substance that falls from heaven, before it becomes contaminated by terrestrial contact (*Nat. Hist.* 11.30-31).[19] The purity of dew is also attested in *Geoponica* 11.18.6, where *drósos*, collected with a feather and applied with a scalpel, is recommended as a cure for ophthalmia. Here dew effects a kind of homeopathic magic, or in Lévi-Strauss' terms, "congruity for therapeutic purposes."[20] One type of moisture, excessive discharge from the eyes,[21] is healed by another, pure liquid distilled from the sky.[22]

Of all extant Greek poets, Euripides is by far the fondest of both "primary" and "developed" uses of *drósos* in contexts of purity (ritual and sexual purity, as well as more mundane cleanliness) and closeness to the gods. In the *Andro-*

----

18 See above, p.44. Recall too that the Muses themselves bestow dew, *Theog.* 81-83.

19 Above, pp.47-48.

20 Lévi-Strauss 1966.9.

21 For this characteristic see LSJ *s.v.* ὀφθαλμία .

22 In other folk traditions too, dew has healing powers. Dew restores sight in Thompson 1932-36, *s.v. Dew*, D1505.5.2 and D1505.5.2.1 (Scandinavia) cf. also D1500.1.18.1, where dew from a saint's grave has healing powers (Ireland). Dundes 1980.116-121 cites various traditions where blindness and other disorders of the eyes are cured by liquids in some ways analogous to dew (e.g. mother's milk: another "water of life").

mache, for example, the heroine faults her mistress Hermione for being sexually jealous of her husband Neoptolemus (who has made Andromache *his* mistress), suggesting that Hermione would be upset if even a "drop of heavenly dew" (ῥανίδ' ὑπαιθρίας δρόσου, 227) should fall on him. Clearly this statement presumes the heavenly purity and innocence of dew, its distance from terrestrial pollutions.

In the *Bacchae*, on the other hand, derivatives of *drósos* appear several times in contexts suggesting the vitality of nature, as experienced by the women in their ecstatic worship of Dionysus. First, the Messenger describes a Theban Bacchant striking a rock with her thyrsus, so that "the dewlike wetness of water gushes out" (δροσώδης ὕδατος ἐκπηδᾶι νοτίς: 705). Later, the chorus of Asiatic Bacchants sing of how they will dance in triumph, "thrusting my neck up to the dewy aether" (δέραν / εἰς αἰθέρα δροσερὸν ῥίπτουσ' : 864-865). In the latter passage, *droserós* 'dewy' is used literally of the aether in which dew is formed, but in the former, *drosṓdēs* 'dewlike' is applied to water springing up from the earth. In both contexts, however, the *connotations* of the *drósos*-derivatives are the same: 'fresh, pure, exuberant'.

Other Euripidean passages show further extensions of meaning for *drósos* and its adjectival derivatives--meanings which are not attested for us prior to Euripides. Springs can be *droserós* 'dewy', as in *Hel.* 1335, *Hipp.* 208, or can themselves be called *drósoi*, *I.A.* 182; *Ion* 96, 117, etc. The water of a river

(*Hel.* 1384, *Hipp.* 78, etc.) is called *drósos* or plural *drósoi*, when qualified by an adjective like *potamîa* 'of a river'. So too with sea-water, when a specifying adjective is used (*pontîa drósos* 'sea-dew' *I.T.* 255, and *thalassîa drósos* 'sea-dew' *I.T.* 1192).[23] Even the water added to wine at a solemn banquet is called *drósos* (*Ion* 1194); so is water which will be used to clean the floor in *Hyps.* fr.1, Col. II.17-18 (Italie) and *Andr.* 167. This anomalous use of *drósos* in Euripides suggests a process of semantic shift, in which the literal meaning of the word becomes less restricted, while its secondary associations stay the same. This probably happens first when adjectives meaning 'dewy, dewlike' are broadly applied to various substances; then the meaning of the root word itself is generalized to mean virtually any kind of fresh water. What began as a metaphorical extension of meaning becomes a new, more extensive "literal" meaning of the word.[24] It appears that Euripides' predilection for forms of *drósos* was remarkable enough to elicit a parody. In his imitations of Euripidean diction in the *Frogs*, Aristophanes twice uses *dros-* words: *drosizómenoi* 'bedewed', of halcyons who have wet their wings in the sea (1312), and *drósos* of the river water which a woman needs to "wash away" a bad dream (1338).

----

23 In this usage Euripides is anticipated by Pindar, who uses ποντίας ἐέρσας , *N.* 7.79, in a difficult metaphor which is discussed below, pp.93-94; and by Aesch. *Eum.* 904-905, ποντίας δρόσου .

24 For a theory of semantic shift, see Waldron 1967.142-161, esp. 149. The same theory figures in a study of Greek metaphor in Silk 1974.29-30.

Despite the very broad semantic range he appropriates for *drósos*, Euripides nevertheless consistently uses the word in a marked way, to mean water with associations of divinity or purity. In the *Ion*, the hero says that he tends Apollo's temple with *drósos*, meaning water from the spring of Castalia (96, 117, 436, and cf. 105-106 *hugraì rhanîdes* 'sacred drops'). *Drósos* is the word for purifying water among the Taurians too: at the beginning of the tragedy herdsmen come to wash their cattle in *enalîa drósos* 'sea-dew' (*I.T.* 255), and later, Orestes and Pylades are supposed to be washed in *thalassîa drósos* 'sea-dew' before being sacrificed to Artemis (1192). (There is a certain irony in Iphigenia's request for sea-water at this point, for far from purifying the "strangers" for sacrifice, she intends to escape with them over the sea.) Similarly, the notion of ritual purification probably underlies Helen's bathing of Menelaus in *potamîa drósos* 'river dew' (*Hel.* 1384): a nuptial bath would be part of the re-creation of their marriage.[25]

Like Ion, Hippolytus is singularly devoted to a god. He offers to Artemis a garland of flowers plucked in a "pure meadow" (*akēratos leimṓn: Hipp.* 73-74, 76-77). No ordinary mortal such as a herdsman or farmer enters there, but (*Hipp.* 77-78):

----

25 *Ros* is also used in Latin to designate pure, fresh water used in rituals: Ovid directs celebrants of the Parilia to wash their hands in "living dew" (*vivo...rore*), *Fasti* 4.778. Cf. also *Aeneid* 6.230, where the tomb of Misenus is sprinkled with 'dew' shaken from olive branches.

μέλισσα λειμῶν' ἠρινὴ διέρχεται,
Αἰδὼς δὲ ποταμίαισι κηπεύει δρόσοις.

the bee in spring traverses the meadow
and Reverence tends it with river dews.

Drósos here is part of a natural-divine world not ordinarily accessible to human beings; only a Hippolytus is worthy to enter the meadow (79-81). Yet like the daughters of Porthaon in Hesiod fr. 26MW, Hippolytus is vainly, or hybristically, rejecting marriage in a setting which itself illustrates the fecundity of nature. He is aware of one aspect of *drósos*, its purity, but does not acknowledge that this same liquid also promotes fertility; and so Hippolytus in the meadow, like the virgin girls among the flowers, remains blind to the cosmological imperative embodied in 'dew'.

But Hippolytus' ideas are not the only perspective on *drósos* in the tragedy, as soon becomes clear. Shortly after his description of the meadow, the women of the chorus also refer to "river dew" (*potamîa drósos*: 126)--this is elevated language for the water where they wash their clothes! For the chorus, however unwittingly, to repeat Hippolytus' phrase in such a mundane sense seems to undercut his reverent notions about 'dew'.

A similar deconstruction of meaning occurs in the next episode. Phaedra in her delirium longs to get a drink of "pure water" (καθαρῶν ὑδάτων: 209) from a "dewy spring" (δροσερᾶς ἀπὸ κρηνῖδος: 208). Not only does Phaedra unwittingly echo the language of Hippolyus, but her yearning for this draft of "dewy water" unconsciously expresses her sexual desire for him.[26] And just as the chorus echoed Hippolytus' "river dews", so now the Nurse diminishes the image of Phaedra's "dewy spring" with her common-sense objection: "Why are you in love with streams of spring water? There's a dewy slope (δροσερὰ κλιτύς) right by the palace where you could get a drink" (225-227).

Exposure. A few contexts in epic and tragedy refer to dew in terms more reminiscent of Columella's dire warnings about it[27] than of the usually favorable connotations of words for 'dew' in Greek poetry. Dew can be a threat to the comfort, health, and even life of men who are unwillingly exposed to nature. In view of such a notion, it may be significant that Athenian youths in military training, who could be expected to spend much time away from the shelter of home, had a special relationship to goddesses whose names and functions connect them with the attributes of dew.[28]

----

26 Glenn 1976, in discussing the erotic symbolism of Phaedra's fantasies here, concludes that "the fountain, the source of flowing liquid, is to be interpreted as a phallic symbol" (p.436), without considering the connotations of 'dew' specifically. See also Segal 1965 for an extended discussion of imagery, especially water imagery, in the *Hippolytus*.

27 E.g. 1.5.8: *concreti rores* threaten the health of plants, animals, and human beings; dew harms individual species of crops in 2.10.29 (vetch), 2.10.10 (beans), 4.19.2 (vines). See above pp. 34-38 for a more extensive discussion of dew in *De Re Rustica* and other Roman writings on agriculture.

28 See below pp.107-112 and 117 for the worship of Agraulus and Pandrosus by the *éphēboi.*

The most striking example of dew as a potentially fatal force is found in Odysseus' soliloquy in *Od.* 5.465-473. Naked, exhausted, lost after his final shipwreck, he finally reaches the shore of Phaeacian Scheria. Wearily he assesses his situation and considers his next move:

> Ὤ μοι ἐγώ, τί πάθω; τί νύ μοι μήκιστα γένηται;
> εἰ μέν κ' ἐν ποταμῶι δυσκηδέα νύκτα φυλάσσω,
> μή μ' ἄμυδις στίβη τε κακὴ καὶ θῆλυς ἐέρση
> ἐξ ὀλιγηπελίης δαμάσηι κεκαφηότα θυμόν·
> αὔρη δ' ἐκ ποταμοῦ ψυχρὴ πνέει ἠῶθι πρό.
> εἰ δέ κεν ἐς κλιτὺν ἀναβὰς καὶ δάσκιον ὕλην
> θάμνοις ἐν πυκινοῖσι καταδράθω, εἴ με μεθήηι
> ῥῖγος καὶ κάματος, γλυκερὸς δέ μοι ὕπνος ἐπέλθηι,
> δείδω μὴ θήρεσσιν ἕλωρ καὶ κύρμα γένωμαι.
>
> Ah me, what do I suffer, what will happen to me at last?
> If I stay at the river through the miserable night,
> evil frost and female dew together
> may conquer my spirit, exhausted in weakness:
> the breeze from the river blows cold before dawn.
> But if I go uphill to the shady wood
> and fall asleep in the dense thickets, even if
> cold and weariness leave me and sweet sleep comes,
> still I fear that I may become prey and plunder for beasts.

Although such a perspective on *eérsē* is unparalleled in epic, Odysseus' fear of exposure is of course quite intelligible, and is specifically confirmed in a number of passages dealing with natural history. Dew and warmth combine to provide the most favorable circumstances for the growth of plants (Theophrastus *C.P.* 3.2.6), but dew and cold evidently provide the worst: even threatening life (Columella 1.5.8). Dew in cold weather is hazardous to animals as different as sheep (*Geoponica* 18.2.7) and crocodiles (Herodotus 2.68.1). Even dew-eating cicadas fall silent in regions where the dewfall is too heavy, without the counter-balancing warmth of the sun (Strabo 6.1.9).

Odysseus himself is already overly wet, swollen with the sea-water which pours out of his mouth and nostrils (*Od.* 5.455-456). He is reduced in fact to infantile status: naked, speechless (456), defenceless, without possessions or identity. In this extremely vulnerable situation even the slight and gentle moisture of dew could make a fatal difference. Yet more is involved than just the difference between wet and dry. In this state Odysseus is completely exposed to the dangers of nature--to wild beasts as well as to cold and dew. His first step in extricating himself from this perilous situation is to *recognize* his opposition to the forces of nature. Then by carrying out typically human activities--talking, assessing, deciding between two dangers, constructing a shelter for himself--he begins to regain his human identity and place in human culture. In such circumstances dew is seen from the point of view of one who is dangerously vulnerable to it: deprived of all attributes of culture, he has temporarily lost his *difference* from the natural world of which dew is a part.

The use of the phrase "female dew" in this context is particularly interesting. As mentioned above,[29] *thêlus* derives from the root **dhē*- 'suckle'. Where the formula occurs elsewhere in archaic poetry, it appears appropriately in a context where dew is food and drink for the cicada (*Sh. Her.* 395, cf. *Anth. Pal.* 6.120.4: *thêlus eérsē* in the same context). Obviously the "female dew" of *Od.* 5.467 has an entirely different connotation. A survey of the uses of *thêlus* and

----

29 P. 43.

thēlúteros in archaic epic shows that, apart from the phrase under consideration (and a single attestation of *thêlus aütē* 'female sound', of the girl's voice that wakes Odysseus from his sleep: *Od.* 6.122), the words are used only of sexually-differentiated beings: goddesses, women, animals.[30] The epithet serves to distinguish the gender 'female' as opposed to male. This could not be the case, however, with *eérsē*, and we must look further to understand the importance of *thêlus eérsē*. In a few instances, *thêlus* appears to be motivated by contextual considerations. Thus only when Hera "deceives Zeus with tricks" and delays the birth of Heracles, is she characterized as *thêlus* (*Il.* 19.97). Again, when Antilochus urges his (male) horses to pass Menelaus' team in the horse race at the funeral games of Patroclus, he warns them that their chief opponent, the mare Aithe, *thêlus eoûsa* 'being female', will shame them if she wins (*Il.* 23.409).

In these cases, *thêlus* signals an anomalous relationship between the sexes, one in which the female is dominant. In the context of Odysseus' soliloquy, *thêlus eérsē*, instead of being gentle, welcome moisture, has the potential to destroy a hero who has just battled and survived the sea itself. To die a victim of exposure on the Phaeacian shore could be an end even more inglorious than death in shipwreck. Yet the dangers of the situation are not entirely unparalleled, for in his wanderings Odysseus has often found himself in the vulnerable

----

30 On the significance of the suffix - τερος as it occurs in θηλύτερος see Benveniste 1948.116-117.

situation of being subject to females or to wild nature.[31] Here "female dew" combines both these elements, with the added irony that the epithet *thêlus*, which otherwise suggests the nurturing characteristics of *eérsē*, is decidedly inappropriate to the context.[32] In short, the use of *thêlus* in this passage suggests that the danger to Odysseus from "female dew" is colored by the epic's complex attitudes toward the feminine in general.[33]

References to the dangers of exposure to dew are not limited to the *Odyssey*. In tragedy the theme recurs several times, especially in connection with the sufferings of Greek soldiers during the long war at Troy. In Sophocles, *Ajax* 1207-1210, for example, the chorus of mariners lament their fate after the hero's suicide:

> . . . κεῖμαι δ' ἀμέριμνος οὕ-
> τως, ἀεὶ πυκιναῖς δρόσοις
> τεγγόμενος κόμας, λυγρᾶς
> μνήματα Τροίας.
>
> So I lie, taking no care, my hair ever wet with heavy dews, memorials of baneful Troy.

----

31 See Taylor 1963 for an extended discussion of Odysseus' encounters with females and the sea as symbolic temptations and threats to his identity.

32 It will be apparent that I believe Homeric epithets are used significantly, whether because they have inherited traditional meaning in connection with a particular theme, as stated most eloquently in Nagy 1979.1-11, or because they are consciously selected for use in a particular context, as posited in Austin 1975.11-80. I am even of the opinion that these two views are not incompatible.

33 On this vast and important subject I will cite as introductory studies only the articles of Arthur 1973 esp. 9-19, Beye 1974, and Foley 1978.

Here dew is connected with the chorus' resentment of the long seige which has removed them from everything protected, familiar and pleasant, and exposed them to physical discomforts of nature as well as the sorrows of war.

In Aeschylus' *Agamemnon* references to *drósos* form a minor pattern of imagery involved with the great polarities of the drama, especially the opposition between nature and culture and between female and male.[34] First, in the prologue the Watchman complains about his "dew-soaked bed" (ἔνδροσον εὐνήν: 12-13) on the palace roof. He is exposed to the discomforts of nature because of Clytemnestra's orders to keep watch for the signal fire from Troy, orders which remove him from the safety and comfort of the house. In the great choral song that follows the Watchman's soliloquy, the old men of Argos fearfully recall how Artemis hated the eagles who killed a pregnant hare with her young--an omen signifying Agamemnon's sacrifice of his daughter Iphigenia as well as his capture of Troy. Ominously they cite the goddess' benevolence toward all young animals, including the *drósoi* 'dew-drops' (141) of fierce lions. Here again 'dew' is part of wild nature, and under the protection of a goddess who opposes the designs of Agamemnon.

In the next episode, as Clytemnestra announces the Greek victory at Troy, in a duplicitous and ironic speech she imagines the pleasures finally enjoyed by the Greek army in a conquered city (*Agam.* 334-337):

----

34 See Zeitlin 1978 for a now classic study of the dichotomies in the *Oresteia*.

ἐν αἰχμαλώτοις Τρωικοῖς οἰκήμασιν
ναίουσιν ἤδη, τῶν ὑπαιθρίων πάγων
δρόσων τ' ἀπαλλαχθέντες· ὡς δ' εὐδαίμονες
ἀφύλακτον εὑδήσουσι πᾶσαν εὐφρόνην.

In captured Trojan houses
now they dwell, escaping aether's springs
and dewfalls (*drósoi*). How happily
will they sleep the whole night long unguarded.

As if echoing her words, but from the perspective of the tedious seige rather than its victorious conclusion, Agamemnon's messenger returning to Argos recalls the constant tribulations of the Trojan War. In language that sounds distinctly non-heroic (like that of the chorus in the *Ajax*) he recounts the physical hardships of sleeping under the open sky for so many years (*Agam.* 560-562):

ἐξ οὐρανοῦ δὲ κἀπὸ γῆς λειμώνιαι
δρόσοι κατεψάκαδον, ἔμπεδον σίνος
ἐσθημάτων τιθέντες ἔνθηρον τρίχα.

From heaven and from earth the meadow
dews dripped down on us, a constant plague
to our clothing, making our hair full of vermin.[35]

In these two speeches as in the Watchman's complaint, *drósos* is a discomfort suffered by men separated from the protections of the house and city, exposed instead to hostile or indifferent nature. In that realm, as the Messenger knows,

----

35 Or "a constant plague, making the fleece of our clothing full of vermin." The use of δρόσοι (always feminine) with both feminine λειμώνιαι and masculine τιθέντες is considered an "intentional solecism" by Denniston and Page 1957.124 on lines 561-562 (following Dover). On the gender of δρόσος, see Schwyzer 1939 (vol. 1).516-517 and 1950 (vol. 2).34 note 1.

dew is generative indeed, but only of vermin.

After the murder of Agamemnon Clytemnestra mentions *drósos* again, but in a very different context. No longer does she pretend empathy with the Greeks at Troy; now her real attitude is revealed. The familiar connotations of *drósos* as fecundating moisture are present, but shockingly perverted, as she describes the murder to the chorus (*Agam.* 1390-1392):

> βάλλει μ' ἐρεμνῆι ψακάδι φοινίας δρόσου
> χαίρουσαν οὐδὲν ἧσσον ἢ διοσδότωι
> γάνει σπορητὸς κάλυκος ἐν λοχεύμασιν.
>
> He strikes me with a dark drop of bloody dew, and I rejoice in it
> no less than a sown field in the Zeus-given[36]
> gleam when birth-pangs come to the grain-sheath.

Agamemnon's blood is *drósos* for his wife--welcome drops that signify the fulfillment of her desires. In comparing herself to a field about to give birth to its harvest, Clytemnestra identifies herself with natural processes,[37] and con-

----

[36] See Fraenkel 1950, vol. 3.655 note 1, for the textual problems. I accept Porson's emendation of διὸς νότωι / γᾶν εἰ. This startling image has not escaped the notice of commentators. In his commentary on 1391f. (p. 656) Fraenkel points out the disparity between the horror of this scene and the joy normally associated with dew and rain, citing *Il.* 23.597ff. and the Aeschylean fragment on the marriage of Sky and Earth (fr. 44N). It should be noted, however, that 'dew' is not an entirely unambivalent image either in epic or in Aeschylus. Dumortier 1935.122-125 catalogues examples of dew and rain imagery in Aeschylus, including some of the passages discussed here (1390-1392, 560-562, 1533-1534), but without developing their importance as a pattern connected with differing points of view in the tragedy.

[37] This mutual identification of woman with grain-field and field with woman provides another example of a widely-attested cultural assumption that the female belongs more closely to "nature" and the male to "culture". See

versely, she personifies the grain-field as a female in labor. For both woman and field, moisture from a male source--Agamemnon or Zeus--facilitates her own purpose. The primordial myth of Sky and Earth is here horribly perverted, for the 'dew' that so delights Clytemnestra is not the fertile seed but the lifeblood of her husband. Clytemnestra's exultation in being sprinkled with bloody *drósos* is the opposite of Odysseus' fear of *eérsē* and his efforts to protect himself from it. She identifies herself with the natural processes of generation and birth, where (male) dew falls welcome on the teeming fields; he sees himself as a potential victim of natural forces, an outsider in the realm of nature, one who must find shelter from the perils of "female dew".

A final comment on Clytemnestra's dew imagery comes shortly after her triumph over the corpse of Agamemnon. The chorus admit their helplessness, and fear for the future of house and city (*Agam.* 1533-1534):

> δέδοικα δ' ὄμβρου κτύπον δομοσφαλῆ
> τὸν αἱματηρόν· ψακὰς δὲ λήγει.
>
> I fear a bloody pelting of rain to bring the palace down;
> the drizzle is giving way to it.

The full effects of Clytemnestra's dewfall of blood have yet to be felt.

Blood. From Clytemnestra's perspective, Agamemnon's "bloody dew" sprinkled her as rain from Zeus sprinkles a fertile field. This comparison is particularly striking because the connotations of the two liquids, blood and rain or dew, are

----

Ortner 1974 for documentation of this idea in a variety of cultures (not including Ancient Greece) and theoretical implications.

so different: in the terms of simile analysis, the link between tenor (here, Agamemnon's blood) and vehicle (rainfall from Zeus) is that of difference, not similarity.[38] But *drósos* links them: it is applied to his blood but is actually more appropriate for Zeus' fertilizing moisture.

Other poets too exploit the contrast between dew and blood, both by referring to blood as dew-like, as in Clytemnestra's speech, and by describing dew as bloody. The former image occurs in Euripides, *I.T.* 442-446, where the chorus of temple maidens, sympathetic to Iphigenia's unhappiness, voice a prayer for vengeance against Helen, the cause of all the dislocations that have resulted from the Trojan War. Knowing that all strangers who come to the Taurian land are sacrificed to Artemis by their mistress Iphigenia, they wish that Helen would land there so that

> with bloody dew (δρόσον αἱματηρὰν) winding around her hair by my mistress' throat-cutting hand, she would die and pay a fitting punishment.

Like Clytemnestra with the blood of Agamemnon, the chorus would find Helen's blood a welcome sight; like Clytemnestra, they call this welcome blood *drósos*.

Homeric epic provides several examples of the "confusion" of blood and dew. In *Il.* 11.52-55, Zeus sends a bloody dewfall as a sign of impending death:

----

38 See Silk 1974.162-3, concerning another image in the same tragedy, *Agam.* 658ff. ("the sea blooming with corpses").

> ... ἐν δὲ κυδοιμὸν
> ὦρσε κακὸν Κρονίδης, κατὰ δ' ὑψόθεν ἧκεν ἐέρσας
> αἵματι μυδαλέας ἐξ αἰθέρος, οὕνεκ' ἔμελλε
> πολλὰς ἰφθίμους κεφαλὰς Ἄιδι προιάψειν.

> ... on them
> Cronus' son drove evil tumult, and down from above he sent
> dew-drops dripping with blood out of the aether, for
> he was going to hurl many mighty heads to Hades.

In the *Iliad*, as we have seen, dewfall normally connotes fertility, nurture, the liquid of life. Thus the softening of Menelaus' anger is compared to *eérsē* "warming" or "cheering" stalks of grain (*Il.* 23.597-600), or dew enhances (albeit ironically) the scene of Zeus' love-making with Hera and subtly suggests his generative powers (*Il.* 14.348, 351).[39]

By contrast, in *Iliad* 11 *eérsē* appears in a dire omen: a dewfall of blood clearly signifies disaster, for it starkly perverts the usual appearance, and therefore the significance, of the celestial drops. Both dew and blood can be considered "vital fluids", two versions of the "liquid of life,"[40] but with important differences. Whereas dew originates in the air and suggests freshness, vitality, and fertilizing moisture, blood on the other hand comes from within the body and is seen only when the body is violated. Its appearance signifies pollution, degeneration, death.

----

[39] The dew simile is analyzed above, p. 32; the Seduction of Zeus is discussed extensively, pp. 17-18 and 54-55.

[40] See p. 10 above, and the references there to Dundes, Onians, and Ninck.

In the structure of this passage describing Zeus' sign to the Achaeans, there are remarkable parallels that suggest an analogy between the dew and the men who will die. Zeus as Sky-Father sends (ἧκεν) down dew-drops from above; likewise, he intends to hurl (προϊάψειν) many men to Hades. Dew-drops, usually a sign of life, now drip with blood; the men, called "many mighty heads" to indicate their strength and vitality,[41] will soon be sent to Hades, lord of the dead. The movement from above to below at the hands of Zeus, and the contrast between the present context and the usual connotations of "dew" and "heads", make the bloody dewfall a grimly appropriate introduction to the carnage of the Great Day of Battle.[42]

A final development of the relationship between dew and blood, and their significance of life or death, appears in the last book of the *Iliad*. Hermes, disguised as a servant of Achilles, guides Priam to Achilles' shelter where he will recover the corpse of Hector, and reassures the old king that despite the abuses of Achilles his son's body remains undefiled (*Il.* 24.416-420):

ἦ μέν μιν περὶ σῆμα ἑοῦ ἑτάροιο φίλοιο

----

41 On the subject of the head as the locus of life, fertility, and strength, Onians cites the birth of Athena from the head of Zeus, the importance of the nod of Zeus (e.g. in *Il.*1.524ff.), the Homeric use of κεφαλός to mean 'life, self' (e.g. in *Il.*18.82), among many other examples. See Onians 1951.95-117.

42 The importance of this image in the structure of the *Iliad* is further signified by its resemblance to the beginning of the epic, where Achilles' wrath "hurled many strong souls to Hades" (πολλὰς δ' ἰφθίμους ψυχὰς Ἄϊδι προΐαψεν: *Il.* 1.3).

ἕλκει ἀκηδέστως, ἠῶς ὅτε δῖα φανήηι,
οὐδέ μιν αἰσχύνει· θηοῖό κεν αὐτὸς ἐπελθὼν
οἷον ἐερσήεις κεῖται, περὶ δ' αἷμα νένιπται,
οὐδέ ποθι μιαρός· σὺν δ' ἕλκεα πάντα μέμυκεν . . .

Indeed around the tomb of his dear friend
he drags him pitilessly, whenever bright dawn appears,
but does not disgrace him; going yourself you would marvel
how dewy he lies, with the blood washed away,
and no pollution at all, but all his wounds are closed. . .

Given the circumstances of his death, his treatment afterwards, and the time that has elapsed since he was killed, it is natural to expect that Hector's corpse would be a gory sight, fitting testimony to Achilles' insatiable hatred. But in fact, thanks to the gods' care for him (*Il.* 24.422-423), he lies fresh and pure, *eersēeis* 'dewy' instead of bloody. His many wounds are closed; the integrity of his body is preserved. Priam, who was worried that Achilles might have torn Hector limb from limb and fed him to the dogs (*Il.* 24.406-409), might have reason to feel encouraged by his guide's description of the state of the corpse. The differences between the expected and the real appearance of Hector at this point are summarized in the contrast between "bloody" and "dewy".

After his moving and successful meeting with Achilles, Priam returns to Troy with the body of Hector, and the women of his family in turn sing their laments over the dead. When her turn comes, the hero's mother Hecabe thinks of all her losses at the hands of Achilles, who now has vainly attempted to revive his friend Patroclus by mutilating Hector's body. With her final words she describes the way her son looks now (*Il.* 24.757-759):

νῦν δέ μοι ἑρσήεις καὶ πρόσφατος ἐν μεγάροισι
κεῖσαι, τῶι ἴκελος ὅν τ' ἀργυρότοξος Ἀπόλλων
οἷς ἀγανοῖσι βέλεσσιν ἐποιχόμενος κατέπεφνεν.

Now before me in the halls, dewy and whole
you lie, like one whom silver-bowed Apollo
with his own gentle arrows has attacked and slain.

Like Hermes and Priam, Hecabe finds that Hector does not look as she might have expected; with *hersēeis* she even echoes Hermes' word *eersēeis* that best describes his fresh and lifelike appearance. She too attributes this remarkable preservation to the care of the gods (*Il.* 24.749-750): they have honored Hector beyond her other sons, for he looks as fresh as if a god had killed him--Apollo with his gentle arrows, which kill swiftly and unexpectedly and do not defile the bodies of the slain. Even the gods cannot restore life to Hector; what they can do they have done, by preserving his corpse intact with the deceptive but perhaps comforting appearance of dewiness.

Chapter Four

DEW/SPEECH/SONG

Greek epic and lyric are self-conscious genres with many ways of referring to themselves and their effects. Water in its various forms, as well as other liquids such as wine and honey, provide some of the most widespread images for language in general and poetry in particular.[1] Metaphors that refer to speech or sound as liquid are extremely common:[2] Nestor's voice "flows sweeter than honey from his tongue" (*Il.* 1.249), the Muses "pour a lament" over the body of Achilles (Pindar, *I.* 8.58), the cicada "pours out its sound" in the heat of summer (*Sh. Her.* 396), or in a particularly interesting figure, a listener who is "parched" with grief (*Theog.* 99) will forget his sorrows when the singer's "sweet voice flows from his mouth" (*Theog.* 97).

As these examples suggest, the affinities between sound and liquid are synaesthetic. Liquids readily pour, spread, flow, soak, fall, and otherwise dif-

----

1 This image apparently existed even in Common Indo-European. See Nagy 1974.229-261, esp. 254-255, for a comparative IE study of the development of the metaphor κλέος (epic fame) = 'stream'.

2 From the point of view of the poet and his audience, many such metaphors may be "dead" unless a perception of their literal sense is stimulated by something in the immediate context, but since I am primarily interested in the *traditional* nature of the metaphor at this point, its "effectiveness" as an image in a particular passage is not at issue. See Silk 1974.27-50 for some interesting points about the difficulty of distinguishing "live" from "dead" metaphors in Greek poetry.

fuse; sounds too are "carried" or "projected" from their source. "Winged words" like other sounds penetrate what would be barriers to sight and touch, to reach even distant places immediately.

'Dew' is used in connection with speech or song (including the "song" of the cicada) in a variety of ways. It may be part of the context in a literal sense or in a simile, or be used metaphorically to represent some other notion. Sometimes *éersē* or *drósos* designates a medium that facilitates or "inspires" the production of sound; sometimes the emphasis is rather on a resemblance between the effects of dew (in the context of natural growth and freshness) and the effects attributed to poetry.in a particular passage. But whatever the context, in all the passages where 'dew' is linked with sound or language, its connotations recall its traditional associations with fertility, celestial purity, or divine origin and transmission (Zeus and his daughters).

First we shall look again at passages in which 'dew' acts to facilitate song or speech, beginning with the song of the *tettix*. The Hesiodic description of a cicada singing on its branch in mid-summer, which was discussed previously in connection with the idea of dew as nurture, suggests that the insect's pleasing sound is related to its diet: "its food and drink are female dew" (*Sh. Her.* 395.[3] Other poems reflect the same idea. An epigram of Leonidas of Tarentum (*Anth.*

----

3 The relevant lines, *Sh. Her.* 393-397, are printed and translated above, pp. 42-43.

Pal. 6.120.3-4) introduces a cicada as "the singer (*aoidós*), tasting a drop of female dew (*hérsē*)." A poem in the collection of lyrics attributed to Anacreon addresses a cicada with a compliment: "Having drunk a little dew (*drósos*) you sing like a king" (*Anacreontea* 32B). This passage recalls the dew-fed kings in the *Theogony*, whom we will soon consider in more detail. Other relevant passages, especially from Hellenistic literature, are cited in E. K. Borthwick's learned article on the diet of poetic "grasshoppers."[4] In addition, Sappho fr. 71.5-8LP, in which unfortunately the beginnings and ends of the lines are missing, preserves a collocation of sweet sounds, dew, and possibly cicadas:

>]μέλ[ος] τι γλύκερον [
>]α μελλιχόφων[
>]δει, λίγυραι δ' ἀη[
>]δροσ[ό]εσσα[

> ...a sweet melody...soft-voiced...clear singers(?)... dewy...

If a form of ἀήδων should be restored in line 7, then Sappho might have been referring to the cicada, which is called ἀήδων in *Anth. Pal.* 7.190, and whose voice is called "clear" (λιγυρήν) in *WD* 583.

Poets compare themselves to cicadas at least as early as Archilochus. In fr. 223W (a paraphrase by Lucian), the iambist apparently warns an enemy that "when a cicada is held by the wing, he sings his loudest"--referring no doubt to Archilochus' own ability to retaliate with invectives even when he seems to be constrained. Such at least is the most plausible interpretation of

----

4 Borthwick 1966.

the fragment, and the one accepted on entomological grounds by L. Bodson in her informative study of cicadas ancient and modern.[5] This however is not the image of the cicada associated with dew; the latter is invariably a sweet-voiced singer, not a noisy, invidious stridulator. The notion of poetry as cicada-sound is illustrated most evocatively in Callimachus' wish to be "the slender, winged one...eating dew (*drósos*), the droplet-food of the divine air" (*Aitia* fr. 1.32-36 Pfeiffer). The image of a small, clear-voiced insect that feeds only on delicate drops of dew evidently appealed to Callimachus' aesthetics.

Like cicadas, poets too have a special relationship to the Muses,[6] and have some hope of an afterlife brighter than oblivion in Hades. Not only does the ethereal diet of cicadas reflect their close relationship to the immortals, especially the Muses,[7] but the passages just cited from Callimachus, Leonidas, the

----

5 Bodson 1976. The author points out that the Archilochus fragment proves that the cicada's production of sound by stridulation rather than by "singing" was known in some quarters much earlier than previously suspected.

6 On the cicadas as friends of the Muses see Plato, *Phaedrus* 259b-c (summarized above, pp. 44-45) and again the cicada poem of Leonidas of Tarentum, *Anth. Pal.* 6.120.7-8: "as much as we are loved by the Muses, so is Athena by us."

7 According to Socrates in the *Phaedrus* passage cited in the preceding note, when cicadas die they report immediately to the Muses. Cicadas, however, could be thought to enjoy a kind of immortality, as suggested by the story that when Eos' lover Tithonus became too old to do anything but chirp, but could not die because he had been made immortal, he was changed into a *téttix* (Scholiast on Lycophron 18). Sappho fr. 55, among others, hints that poets can enjoy a more blessed afterlife than mortals who "have no share in the roses of Pieria".

Anacreontea and the *Shield of Heracles* would also suggest that their clear song depends in some way on the purity of their diet. In the charming image of *Anacreontea* 32B,

> ὀλίγην δρόσον πεπωκώς
> βασιλεὺς ὅπως ἀείδεις
>
> having drunk a little dew, you sing like a king.

there may even be a play on the picture of the cicada as a miniature bard, who like Demodocus in *Od.* 8.70 for example, has his drink beside him as he starts to sing.

Another kind of relationship between 'dew' and sound, this time having to do with human speech, is attested in *Theog.* 81-93, the section of the prooemium that deals with the relationship between Muses and kings. A great deal has been written about this important and problematic passage recently,[8] and I will

----

[8] Some important recent discussions: Puelma 1972: emphasis on the differences, even dichotomy, between singer and king, based on the fable of the hawk and the nightingale in the *Works and Days*. Roth 1976: focus on the sociological role of "kings" in Hesiodic and earlier Indo-European society, suggesting that the Muses are important because kings were primarily judicial administrators, for whom it was essential to memorize oral law-codes. Pucci 1977.17-21: an important illustration of his thesis that in Hesiod the concepts of truth and "straight speech" are highly ambivalent: the king's "straight judgments", for example, are framed in "devious words"; his language is no more "straight" than the poet's. Duban 1980: a comprehensive study of the relationships between Muses and kings that takes careful account of previous scholarship on many aspects of the problem, proposing that for Hesiod it was original but "inevitable" (11) to connect the Muses with kings as well as poets, because kings too must use language effectively. This last point especially concurs with my own conclusions. Arthur 1983.109-111: based on Pucci's discussion of the devious language of both king and bard, with emphasis on the sexual and alimentary connotations of

not attempt to cover all aspects of it, but will focus instead on a few topics that pertain to our understanding of *eérsē* or *drósos* and their affinities with language--a topic which has not yet been sufficiently considered by the existing commentaries on the passage.

First, the Muses grant not only kings but also singers the ability to speak with a special language, effective beyond ordinary speech. When the Muses appear to the shepherd Hesiod on Helicon (*Theog*. 22-34), claiming that they can sing either the truth or lies like the truth, they give him a laurel sceptre, breathe into him a divine voice (ἐνέπνευσαν δέ μοι αὐδὴν θέσπιν, 31-32) so that he may sing (κλείοιμι, 32) past and future things, and order him to hymn the race of the gods. Some lines later the *Theogony* describes the effects of such a song: the singer's voice flows sweet (*glukerḗ*) from his mouth. Whenever he sings the deeds of gods or of heroes, his audience forgets present sorrows, for the gifts of the Muses turn them away (*parétrape*) from grief (97-103). In short, the breath of the Muses inspires sweet song in their servant (*therápōn:* 100), and his voice diverts those who hear him.[9]

The Muses give sweet speech to kings in a different way and with somewhat different results. When they honor a king it is at his birth (*Theog*. 82), in contrast to their appearance before the shepherd/singer on Helicon. Instead of

----

9 For a more detailed analysis of the relationship between Muse and king in early Greece, see Detienne 1973.3-27.

breathing into him, they pour sweet dew (*glukerḕ eérsē:* 83) on his tongue,[10] and honeyed words (*épea meîlicha:* 84)[11] flow from his mouth. For both king and singer then, an encounter with the Muses results in sweet speech of a sort, yet there may be a correlation between the different modes of their "inspiration" and the different effects of their speech. Whereas the singer's voice "turns *aside*" (*parétrape:* 103) the mind of an individual[12] hearer from his private grief, kings[13] on the other hand, even though ambiguously "persuading" (*paraiphámenoi*, literally 'talking *aside*': 90),[14] nevertheless perform "works that turn *back*" (*metátropa érga*: 89) for the benefit of injured or wronged people (plural λαοῖς : 88, cf. 84-85).

----

10 Arthur 1983.111 connects this with bees' methods of depositing honey in their cells, and with their method of reproduction as well; I believe that 'dew' is significant here in its own right, not just as a word for "honey", so I cannot fully agree with this interpretation of *eérsē*, nor with the assumption in Roth 1976.332 that the *eérsē* here is "honey or mead". Nowhere in archaic poetry is honey called *eérsē*, despite the likelihood (based on this passage) that already in this early period bees were thought to make honey from dew.

11 Although the etymology of μείλιχος and its relationship, if any, to μέλι is not certain, it is nevertheless generally and correctly assumed that the two words were perceived as being connected in popular etymology. Cf. Chantraine 1968-77, *s.v.* μειλία.

12 Note singular τις and verbs: 98-99, 101-102.

13 In the plural, 88-90: as a class, perhaps in contrast to the individual singer, 100.

14 Here I follow closely the analysis of Pucci 1977.17-18, although I do not agree entirely with his conclusions on the degree of ambivalence in the king's discourse.

There is a difference here, hard as it is to define. Poets are literally "inspired", breathed on by the Muses, if the experience of Hesiod on Helicon, recounted in such dramatic narrative fashion, is paradigmatic. They are not "born with" their gift, but it comes to them later in life. When the Muses come to Hesiod, they make what sounds like a teasing remark about their ability to speak truth or convincing falsehood at will (*Theog.* 27-28): is this also the nature of the poet's discourse? When he diverts his audience, making each one forget the present situation (102-103), surely his function--as described in this context, at least--is neither to tell the truth nor to lie, but to bring about an effacement of present cares. His effect is primarily psychological, and since no change is made in the external situation, probably temporary.

The speech of a king properly has a different purpose: to persuade his people to return what was wrongly taken, to turn back to straight justice. His role is primarily social, and the effects of his words should be more permanent, for the circumstances which brought about the present troubles will be changed if he is truly persuasive. The king may receive this gift at the moment of his birth (*Theog.* 81-82): his sweet speech is not the result of a dramatic personal inspiration, but a capacity "natural" to him if he is truly an honored king. Kings are of Zeus (96), Zeus-nurtured (82), and the *eérsē* that sweetens their speech is dispensed at will by the "daughters of great Zeus" (81). The relationship to Zeus, the suggestion of an inherent capacity for kingship, and the results of the king's speech which are more tangible and less mysterious than the results of the singer's song--to these characteristics may be related also the fact that a king's sweet words come from tactile, liquid 'dew' rather than the invisible, evanescent 'breath' of the Muses. But there is more: the "honeyed words" (*épea meîlicha*: 84) of a king win him "honeyed reverence" (*aidṑs meilíchios*: 92) from his people. Dew from the Muses thus produces a doubly sweet result: social harmony for the people, and popular respect for the just, persuasive king.

Besides being regarded as a source of song and speech, 'dew' (both *eérsē* and *drósos*) is developed in Greek poetry as an important image of what song does. Nowhere else is this figure exploited as it is in Pindar. Reference to the function of the poet and the benefits of his song is of course an important part of epinician convention, and Pindar uses many images for his praise-poetry and the blessings it confers.[15] Water is among the most important of these images, as anyone familiar with the opening lines of *Olympian* 1 is aware.[16] In Pindar, the metaphorical 'water' of song or praise does many of the same good things that real water can do. Song and the glory it brings may itself be considered a stream (*O*. 14.12, *N*. 7.12 and 62, *I*. 7.19). It can be poured (*P*. 1.8, *I*. 8.58), distilled (*P*. 4.137), or sprinkled (*P*. 8.57, *I*. 6.21) as a libation. The poet can

----

15 See Simpson 1969.437-438 note 1 for bibliography on the subject of Pindaric images of song.

16 Gianotti 1975.110-111 discusses briefly some relevant aspects of water symbolism in Pindar, including its associations with prophecy, immortality, and purification. Gianotti concludes (111): "apparirà chiaro...che Pindaro si vale dell'immagine dell'acqua per indicare aspetti propri del suo canto, anche là dove recupera vecchie concezioni." It is with such *vecchie concezioni* that I am particularly concerned.

give a drink of song to the thirsty athlete (*I.* 6.74, *N.* 3.76-79) and his songs can water the victor's land (*I.* 6.64, cf. also fr. 6.b.f). A song of praise heals weary limbs better than warm water (*N.* 4.4).

Uniformly in Pindar as usually in Homeric epic, however, *ómbros* 'rain' is harmful or unpleasant (*P.* 4.81, 5.10, 6.10). It even serves as a metaphor for the Persian invasion (*I.* 5.49.)[17] *Éersa* and *drósos,*[18] in contrast, have good connotations: in each of their six occurrences in Pindar they refer directly or obliquely to the benefits conferred by praise-poetry.

In two odes *drósos* or *éersa* is offered as a drink. The celebrated prelude of *Olympian* 7 describes a "cup foaming with the dew of the vine" (Φιάλαν . . . ἀμπέλου καχλάζοισαν δρόσωι : 1-2) which is handed from father of the bride to bridegroom at a wedding feast, thereby linking the two families in a harmonious and productive union (4-6). This wedding cup is compared to the gift Pindar himself presents to Olympian and Pythian victors: "liquid nectar, gift of the Muses...sweet fruit of the mind" (νέκταρ χυτόν, Μοισᾶν δόσιν . . . γλυκὺν καρπὸν φρενός : 7-8). The cup of wine is a rich symbol in itself, as D. Young observes.[19] Especially within the context of a

----

[17] But the derivative *ómbrios* 'rainy' connotes a neutral or even welcome shower in *O.* 11.3.

[18] Pindar and Sappho, it will be recalled, are the only archaic poets in whose works both of the primary words for 'dew' are attested. In Pindar *éersa* occurs in *N.* 2, 3 and 7; *drósos* in *O.* 7, *P.* 5 and *I.* 6.

wedding feast, to call the wine *drósos* further develops its meaning, for this word of course has its own set of appropriate connotations: 'dew' is vital, fertile moisture (recall Pliny's *genitalis ros* and Nonnus' γαμίη ἐέρση) likely to suggest the fertile generation of life that the marriage will produce. Pindar's gift of the nectar of song is like the wedding cup in that it too bestows and celebrates the vitality and blessings appropriate to those who have achieved victory in the great games. Moreover, following Young's interpretation of the passage, we note that the source of the wine-dew is suggested in the opening line: the bride's father himself receives the cup from a "wealthy hand" (1) before he gives it to the bridegroom. Corresponding to this part of the simile is the statement in the tenor of the passage that Pindar receives from the Muses the "liquid nectar" he will in turn present to the athletes: the source of his song is not forgotten.

Another drink of 'dew' is described in *Nemean* 3, this time presented to the athlete from the poet. Here then the connection between 'dew' and poetry is not simply suggested in a simile as it was in the poem just discussed, but is part of an important theme of the poem as a whole. The ode begins with the statement that "different deeds thirst for different things" (διψῆι δὲ πρᾶγος ἄλλο μὲν ἄλλου : 6), and ends with Pindar's claim that he is sending the celebrant a "drink of song" (76-79):

----

[19] Young 1968.73-76. Young does not comment on the use of *drósos*.

> ... χαῖρε, φίλος· ἐγὼ τόδε τοι
> πέμπω μεμιγμένον μέλι λευκῶι
> σὺν γάλακτι, κιρναμένα δ' ἔερσ' ἀμφέπει,
> πόμ' ἀοίδιμον ...
>
> Hail, friend! I send you this honey mingled with white milk, and dew mixed in surrounds it, a drink of song...

Thus the poem is framed by the image of a thirst which is satisfied by song. A similar figure seems to underlie a passage in the *Theogony*. There, the singer's voice "flows" (ῥέει: 97), in the conventional and probably "dead" metaphor. But the metaphor comes to life when we hear that the singer's audience is "parched" (ἄζηται: 99) with troubles, which are forgotten when he hears the song. I would even suggest that a similar surcease of discomfort is reflected in another Hesiodic passage, where the dew-eating cicada "pours" (χέει) forth its voice in the season when Sirius "parches" (ἄζει) the skin (*Sh. Her.* 396-397).

The ingredients[20] of Pindar's "drink of song" have their own significance as well. Honey is of course a particularly rich symbol, connotating the preservation of life as well as the sweetness of song.[21] Milk and honey together

----

20 Norwood 1945.170-171 brings some practical experience to the question of whether the drink really consisted of three components or only of two, milk and honey, with *éersa* meaning the foam produced when they were mixed together--an interpretation which Norwood says prevailed among all earlier commentators. Norwood himself tried mixing milk and honey together, and discovered that no foam at all was produced; he concludes that the 'dew' is really a separate ingredient along with milk and honey!

21 See Duchemin 1955.42-43 for a useful review of relevant material on honey in Greek. Chiefly on this basis she concludes that Pindar's drink here is a "breuvage d'immortalité" (43).

reflect the simple, unfermented ingredients required in certain rituals, especially to gods of the earth.[22] They also suggest the effortless food-gathering of the Golden Age,[23] a significance that may also be shared with 'dew', as suggested earlier in connection with epic contexts involving the role of dew and rain in agriculture.[24] *Éersa* also reflects the purity of the drink, and the blessings of abundant life that are freely bestowed by celestial moisture. In sum, Pindar's "drink" is a beverage well calculated to satisfy the athlete's thirst for song by connecting him to the life of the gods.

'Dew' is again related to the poet's function in *Nemean* 8.40. Here it appears not as a drink but in its familiar role as fertilizing force in nature, in a simile comparing vegetal growth to the enhancement of *aretá* 'excellence, achievement' (40-43):

> *αὔξεται δ' ἀρετὰ χλωραῖς ἐέρσαις*
> *ὡς ὅτε δένδρεον· ἀίσσει*
> *ἐν σοφοῖς ἀνδρῶν ἀερθεῖσ' ἐν δικαίοις τε πρὸς ὑγρὸν αἰθέρα.*
>
> *Aretá* grows like a tree with fresh dews. Lifted among wise and just men it shoots up toward the moist aether.

Without mentioning the "fresh dews" specifically, C. Carey has shown that this

----

22 See *SIG* 1025.34 (Kos) and Eur. *Or.* 115. For an important discussion of milk, honey and wine in Greek ritual see Graf 1980.

23 See Pucci 1977.21 for the role of honey in "Golden Age" contexts.

24 See above, pp.31-32, 41-42.

poem deals repeatedly with images of growth and decay.[25] "Fresh dews" are part of this pattern, clearly indicating the tree's healthy growth, reminiscent of the "flourishing dew" that encourages the abundance of grain and vines on Ithaca (*Od.* 13.244-245). Correspondingly, the poet's song will encourage the development of the athlete's prowess and reputation, his *aretá*.[26] In the proper environment, among wise and just men, the achievement nurtured as if by *éersa* from the sky will ascend in full circle to the "moist aether".

In one of his most complex odes, *Nemean* 7, Pindar contrasts the task of weaving garlands of song with the Muse's enduring creation of an elaborate threefold work, made of gold, ivory and a third component called a "delicate flower plucked from the sea-dew" (λείριον ἄνθεμον ποντίας ὑφελ-οῖσ' ἐέρσας : 79). On the basis of a suggestion in the scholia, this "flower" is usually understood to be coral, which was thought to harden only upon contact with air. Yet the evidence of archaeology suggests that the use of coral was scarcely known to the Greeks of Pindar's time.[27] The same scholion suggests another possible referent, sea-purple dye; this possibility is taken up and

----

25 Carey 1976, whose suggestions I follow (p. 35 and note 47) for the punctuation and interpretation of this passage, especially ἀίσσει. See also Irwin 1974.31-78 esp. 52 and 77, for interesting comments on the uses of χλωρός in this passage and others.

26 On this double inference of the word see Slater 1969, *s.v.* ἀρετά.

27 For what evidence there is see Petegorsky 1982.150 and note 17.

argued in great detail in D. Petegorsky's recent dissertation.[28] In this interpretation, the Muse's creation is not a diadem of gold, ivory and coral glued together, but a woven headband of cloth dyed purple and ornamented with gold and ivory. In an illuminating earlier essay on this ode,[29] C. P. Segal (who accepts the usual interpretion of the "flower" as coral) argues that each element in the Muse's "jewel" is concerned with major themes of the poem, especially permanence, light, and growth. Petegorsky establishes that both "dew" and "flower", and sea-purple as well, are traditionally associated with the idea of song, which is itself often designated as a woven thing. For Segal too, 'dew' is evocative of "poetry and the capacity of men to reach what is creative and life-giving in their world".[30] With this I would concur, emphasizing however that the 'dew' is not merely human or terrestrial in origin: in several of the passages specific goddesses, such as the Muse in *Nemean* 7, are closely linked with the *drósos* or *éersa* that enhances the poet's song.

----

28 Petegorsky 1982.141-174, esp. 160-165 on 'dew'. I had the opportunity to discuss some aspects of *éersa* in this passage with the author, and to an extent our ideas are mutually dependent. I am happy to record my gratitude for this helpful and stimulating exchange.

29 Segal 1967, a study which greatly improved on earlier allegorical analyses of the components of the "diadem": Norwood 1945.107-109, Finley 1955.102.

30 Segal 1967.466; this conclusion follows the author's survey of the uses of 'dew' in *N*.3 and 8 and *I*.6, which is the most complete study of the image in Pindar I have seen.

In the two remaining attestations of dew in Pindar, *drósos* is equated with praise-poetry as it benefits not the victor alone, but his family as well. *Isthmian* 6 abounds in images of song as liquid, beginning with the "mixing-bowl of the Muses" from which libations are poured over Aegina as offerings for victories in the games (1-9). In the central myth of the ode, Heracles pours a libation of nectar from a golden cup as he prays for the birth of a son to his guest-friend Telamon (39-46). Returning to the theme of victory and celebration, Pindar enumerates the past accomplishments of the sons of Lampon and their maternal uncle (62-66):

> ...ἀνὰ δ' ἄγαγον ἐς φάος οἵαν μοῖραν ὕμνων·
> τὰν Ψαλυχιαδᾶν δὲ πάτραν Χαρίτων
> ἄρδοντι καλλίστᾳ δρόσῳ,
> τόν τε Θεμιστίου ὀρθώσαντες οἶκον τάνδε πόλιν
> θεοφιλῆ ναίοισι·
>
> What an allotment of songs they have brought to light! They water the clan of the Psalychiadae with the loveliest dew of the Charites, and setting straight the house of Themistius they dwell in this city dear to the gods.

As often in Pindar, the Charites preside over victory celebrations here; their *drósos* can almost be equated with the songs themselves.[31] In an image drawn from agriculture, the young athletes are said to water their *pátra* 'clan, fatherland' with this 'dew', so that the fame of their victories, as brought to light by songs,[32] may cause their whole family to flourish.

----

31 Similarities between the Muses and the Charites in Pindar are developed in Duchemin 1955.54ff.

32 Cf. the similar image in Pindar fr. 6.b.f: ἄρδοντ' ἀοιδαῖς 'water with

Finally, in *Pythian* 5, written for the chariot victory of Arcesilas IV of Cyrene, Pindar again speaks of the celebrations as *drósos* (96-103):

> ἄτερθε δὲ πρὸ δωμάτων ἕτεροι λαχόντες Ἀίδαν
> βασιλέες ἱεροί
> ἐντί· μεγαλᾶν δ' ἀρετᾶν
> δρόσωι μαλθακᾶι
> ῥανθεισᾶν κώμων ὑπὸ χεύμασιν,
> ἀκούοντί που χθονίαι φρενί,
> σφὸν ὄλβον υἱῶι τε κοινὰν χάριν
> ἔνδικον τ' Ἀρκεσίλαι . . .

> Apart, in front of the palace, are other holy kings who have their lot in Hades; somehow with chthonic mind they hear the great *aretá*, sprinkled with gentle dew by the streams of the celebrations--their own happiness and the well-deserved grace they share with Arcesilas.

In other odes too Pindar imagines how the victory is being communicated to dead members of the athlete's family.[33] In *Nemean* 4.85-86, the victor's dead uncle will hear the poet's tongue; in *Olympian* 14.20-24. Echo is sent to tell a dead father of his son's triumph; in *Olympian* 8.77-84, Angelia ('Messenger') the daughter of Hermes will announce the victory to the celebrant's father in Hades, where he may in turn share the good news with his own brother there, for "even to the dead is there some share of deeds."

So in *Pythian* 5 the achievement of Arcesilas is 'bedewed' by the celebrations in his honor, with the result that it can be *heard* by his royal ancestors. The gentle penetrating moisture of dew is an appropriate metaphor for this

----

songs'.

[33] See now C. Segal 1985 on "Messages to the Underworld" in Pindaric poetry.

subtle communication and further suggests that the dead, who are often envisioned as thirsty,[34] may be refreshed and nurtured by the news.

In other genres of poetry too 'dew' sometimes communicates with or enhances the world of the dead. The image of Acheron's "dewy banks" in Sappho fr. 95LP, for example, lightens the usually somber atmosphere of Hades.[35] Closer in theme to Pindar's use of 'dew' in *Pythian* 5, however, is an epitaph for Anacreon, attributed to Simonides, which addresses the grapevine growing over the poet's grave (Simonides 125.9-10D):

> *καί μιν ἀεὶ τέγγοι νοτερὴ δρόσος, ἧς ὁ γεραιός*
> *λαρότερον μαλακῶν ἔπνεεν ἐκ στομάτων.*
>
> and may moist dew ever drench him, which the old man
> breathed forth so sweetly from his gentle mouth.

Here *drósos* implies the "dew of the vine" (cf. *O.* 7.2), alluding to Anacreon's famous fondness for wine.[36] But 'dew' also represents Anacreon's poetry ("which the old man breathed forth so sweetly") and may suggest especially its erotic, youthful, lively subject matter. Most important, 'dew' is imagined to filter down to the poet buried beneath the vine, to provide him with a taste of the blessings he delighted in during his life.[37] Again, 'dew' is not just a fluid

----

[34] See Kurtz and Boardman 1971.209.

[35] Boedeker 1979. See also above, p. 60, for a summary of the conclusions reached there.

[36] A statue of Anacreon on the Athenian Acropolis, for example, showed him singing "in his cups" (*ἐν μέθηι*) : Paus. 1.25.1.

medium of communication from the world above to that below, but a welcome source of something like life.

The relationships we have considered between 'dew' and language may illuminate the nature of both. Dew is part of that natural-divine world in which mortals may sometimes participate, especially associated with Zeus and his daughters the Muses, who bestow persuasive speech in the form of sweet dew placed in the mouths of kings. 'Dew' is also refreshing, transformative moisture: for the cicada, and for the poet who would identify himself with that clear-voiced insect, it enables him to pour forth his sweet song even in parched surroundings. The fertilizing and nurturing attributes of dew are further confirmed in Pindar's images of song and vegetal life, through both of which 'dew' promotes growth and fulfilment.

Analogies with dew assimilate language to the realm of physical nature, an aspect of poetry rightly emphasized by Detienne.[38] Speech and song can encourage life, growth, healing, and can penetrate even to the underworld. The powers of language are emphasized especially in praise-poetry, where the 'dew' of celebratory song, ultimately a gift of the Muses or Charites, is poured out by

----

37 Certain Greek funeral practices illustrate the idea of giving drink to the dead, as reported in Kurtz and Boardman 1971. Cups and water jugs are often found in burials, and in cremations the urn for the ashes was itself often a hydria (pp. 209-210). In addition, vine tendrils have been identified among Geometric grave offerings in Attica and elsewhere (p. 64). See Alexiou 1974.202-204 for Modern Greek references to the "thirsty dead".

38 See the important discussion of this point in Detienne 1973.55.

the poet to increase the fame and prosperity of the victor, his family, and his fatherland. Like the Hesiodic king, the praise poet in Pindar's scheme is an instrument through which the blessings of 'dew' are shed on those who hear him.

Chapter Five

DEW AND ATHENIAN AUTOCHTHONY

Many of the themes connected with *drósos* and *eérsē* in Greek literature come together in a complex of myths and rituals centered on the Athenian Acropolis. On the one hand there is the myth of the birth of Erichthonius and his exposure by the daughters of Cecrops, where we find not only the names of the Cecropids *Pándrosos* 'All-Dew(y)' and *Hérsē* 'Dew', but also the motifs of generation from the 'dew' of the father, birth from fecundated earth, nurture and subsequently exposure at the hands of females figures connected with 'dew', and possibly communication from above to below. On the other hand is the secret rite performed by the Arrhephoroi, a rite whose name *Arrhēphorîa* is often interpreted as 'Dew-Bearing', and whose young ministrants have been shown to have a close affinity to the Cecropids in the myth.[1] After examining that affinity and

----

1 This topic has a noble place in the history of scholarship on Greek religion, and the literature on it is vast. I would point out two works for their comprehensive presentation of source material: Powell 1906 with its useful collection of the ancient testimony, and Burkert 1966 where modern scholarship is extensively discussed. Although I disagree with Burker's interpretation at several points, I admire his magisterial re-reading of the Arrhephoria in a sociological context, as opposed to earlier explanations of the rite as a "fertility festival"--an interpretation which now appears unnecessarily reductionist. Since the appearance of Burkert's article the intensity of discussion has increased; for my purposes, the most important recent works that deal extensively with the subject are: Schmidt 1968, Astour 1969, Brelich 1969.229-302, Burkert 1972.169-173, Kron 1976.55-72, Martin and Metzger 1976.173-174, Metzger 1976, Burkert 1977.347-349, Calame 1977 vol. 1.236-241, Peradotto 1977, Chirassi-Colombo 1979 esp. 47-48, Loraux

the name of the ritual, this chapter will discuss several aspects of both myth and rite from the perspective of what we have seen about the functions of 'dew' in the preceding chapters.

First the rite and its participants. Each year four young Athenian girls were selected as Arrhephoroi by the Archon Basileus. Pausanias 1.27.3 describes an unnamed nocturnal rite, undoubtedly the ritual called "Arrhephoria" in other sources, in which two of the Arrhephoroi went from a point near the precinct of Pandrosus on the north side of the Acropolis via an underground path to an enclosure below: excavations make it clear that this path led to, or through, the cave of Aglaurus on the steep north slope of the hill, immediately beneath the House of the Arrhephoroi.[2] Inscriptions dating from the third century B.C. and later honor a number of girls who "served as *errēphóros* (occasionally *arrēphóros* in later inscriptions) to Athena Polias and Pandrosus".[3] The patronage of these two figures strongly suggests that the Arrhephoroi mentioned in Pausanias have some connection with the Cecropids and the birth of Erichthonius. Besides their mysterious journey under the Acropolis, the Arrhephoroi also began the weaving of Athena's woolen peplos every year; cor-

----

1979=Loraux 1981.35-73, Kadletz 1982, Zeitlin 1982 esp. 150-153, and Simon 1983.39-46.

2 See Travlos 1971.70-71 (fig. 91) for a convenient overview of these sites in relation to each other and to the rest of the Acropolis; pp. 72-75 for succinct discussion of the cave of Aglauros and its spring; and pp. 213-218 for a description and plans of the Erechtheum including the temple of Pandrosus.

3 See Burkert 1966.5-6.

responding to this are the traditions that Pandrosus with her sisters introduced the wearing of wool clothing (Photius *s.v.* προτόνιον) and that Aglaurus was the first to adorn the statues of gods (Bekker, *Anec. Gr.* i.270). Thus in this connection too the Arrhephoroi resemble the Cecropids.

The two groups of girls may likewise share similar names: *Hérsē* and *Pan-drósos* evidently mean 'Dew' and 'All-Dew(y)'; *Arrhē-* or *Errhē-* or *Hersē-phóros* has been interpreted etymologically as 'Dew-Bearer' by some scholars from ancient times to the present, although the etymology is far from clear and is regarded as uncertain by both Frisk and Chantraine.[4] Even the form of the word itself is variable. We do not know how Athenians of the archaic and classical periods pronounced the first element of the compound, but *errhē* is the most likely conjecture: the earliest attestation of a form of this compound is in Aristophanes, where it unfortunately occurs as a verb with temporal augment (*ērrhē-phóroun*, *Lys.* 641) which could derive from either *errhē-* or *arrhē-*. A scholion on this passage presents two possible etymologies which remain the most frequently discussed: if the name of the festival begins with *a*, then it is a shortened form of *arrhēto-phorîa* 'bearing unmentionable things'; if it begins with *(h)e*, then it signifies *(h)ersē-phorîa* 'dew-bearing', "because they celebrate it for (H)erse the daughter of Cecrops." No source is cited for the former conjecture, but the latter ('dew-bearing') is attributed to Istros, a student of Athe-

----

[4] Frisk 1960-1970 and Chantraine 1968-77, *s.v.* ἀρρηφόρος.

nian antiquities who lived in the third century B.C. Virtually identical information is included in a number of ancient compilations (Moeris the Atticist, Hesychius, the *Suda*, etc.), but unlike the scholiast on Aristophanes they do not name the source of the proposed interpretation.[5] Fick, Forbes, and Burkert have plausibly conjectured that the first elements of *arrhē-phorîa* derives ultimately from **wers-*, the root of *hérsē*.[6] Nevertheless the existence of the multiforms and their multiple interpretations is instructive: if 'dew-bearing' was the original name, this was generally forgotten although it remained a plausible conjecture to scholars such as Istros. The form *arrhē-phorîa* probably became assimilated in the popular imagination with *arrhēto-phorîa* 'bearing unmentionable things', which shows that the connection with 'dew' was no longer under-

----

[5] See Powell 1906.79-80 for the testimony.

[6] See Burkert 1966.17 note 1 for a sympathetic treatment of Fick's etymology of ἀρρη-φόρος from ἕρση 'dew' (Fick 1910.132-133); Burkert himself proposes that ἀρρη- derives from the zero-grade of **wers-*, namely **wr̥sā* > *ἄρσα > *ἄρση > ἄρρη. Forbes 1958.255-256 suggests virtually the same possibility for the derivation of ἀρρη-, but thinks it more likely that later ἀρρη- was dissimilated from earlier ἐρρη- by popular connection with ἄρρην 'male'. Her proposal was rejected summarily by Frisk 1960 and Chantraine 1968-77 *s.v.* ἀρρηφόρος , but given the chronology of ἐρρη- > ἀρρη- in the inscriptions, and the testimony of Istros (contemporary with the ἐρρη- inscriptions), the sequence ἐρση- > ἐρρη- > ἀρρη- does not seem unlikely. In Attic it is normal that the unaspirated forms preserve no trace of initial *w*. For an apparent Attic variant of Homeric ἐρσήεντα, see Hesychius' gloss ἐρρήεντα· δροσώδη, καταψυκτικά. The derivation of ἀρρη-φορία from ἀρρητο-φορία, on the other hand, as Chantraine notes, offers no explanation for the loss of the syllable -το-; Chantraine also tends to reject the proposal of Adrados 1951 that ἐρση- and ἀρρη- are dialect variants, derived from ἔρσην / ἄρσην 'male': in compounds ἐρσενο- / ἀρσενο- would be expected.

stood.[7] Pausanias (second century B.C.) after all candidly admits that the rite was a mystery to him and quite unknown by most (1.27.3).

As for the three Cecropids, all their names too seem to suggest 'water' in some form. The name *Hérsē* 'dew', with its unassimilated *-rs-*, cannot be Attic; Frisk suggests that it may be a "hieratic Ionism",[8] and indeed a Herse was worshipped at Ionian Erythrae on the coast of Asia Minor.[9] Even if we do not follow J. Harrison in reducing Herse to "a mere etymological eponymous of the festival Hersephoria",[10] still it appears that she alone of the Cecropids has no cult place in Athens, and that Attic women swore by Pandrosus and Aglaurus but never by Herse (according to the scholiast on Aristoph. *Thesm.* 533), suggesting that Herse is a later addition to the tradition. Perhaps her name was added to the Cecropids' because of its analogy to Pandrosus 'All-Dew(y)'--or even (as the comment from Istros would suggest) because it resembles the name of the rite.

Pandrosus, on the other hand, is well established on the Acropolis with her precinct just west of the Erechtheum, and is closely associated with the cult of

----

7 This is also the conclusion of Nilsson 1967.441.

8 Frisk 1960, *s.v.* ἕρση.

9 Engelmann and Merkelbach 1973, vol. 2 #206 (pp. 342-343).

10 Harrison 1891.351.

Athena Polias.[11] The Athenian sacred olive tree is located in the Pandroseum; Burkert finds in this an analogy with Minoan-Mycenaean tree-goddesses indicating the great antiquity of Pandrosus' cult at Athens.[12] At any rate her relationship with the olive further suggests that the figure of Pandrosus is connected with fertilizing moisture.[13]

Aglaurus too is associated with water: her cave on the north slope of the Acropolis contained a deep spring. The spring was used only briefly in Mycenaean times, but its existence may have influenced this variant of her name (sometimes attested as *Agraulos*), for *ágl-auros* may be interpreted as a compound of *aglaós* 'bright' and a very archaic word *aúrā̆* 'water'.[14]

One of the stories about Aglaurus and her sisters has even been interpreted as a reflection of rain-invoking rituals. The story involves a recurring pattern in Athenian mythology: the sacrifice of a maiden or group of maidens saves the city from disaster. During the war between Athens and Eleusis, an oracle pre-

----

11 E.g. the two are cited together on a number of the dedicatory inscriptions of Arrhephoroi (*I.G.* II/III^2 3472, 3488, 3515). The epheboi sacrifice to them jointly (together with Kourotrophos: *I.G.* II/III^2 1039). The sacrifice of an ox to Athena necessitates the sacrifice of a sheep to Pandrosus (Philochorus in Harpocration, *s.v.* ἐπίβοιον). Athena is sometimes called Athena Pandrosus (scholiast on Aristoph. *Lys.*439).

12 Burkert 1977.93.

13 Simon 1983.45-46 particularly emphasizes the importance of the olive tree in connection with the "dew-sisters" and the rite of Arrhephoria.

14 See Chantraine 1968 *s.v.* ἄναυρος.

dicted victory for Athens upon an act of self-sacrifice for the city; Aglaurus willingly threw herself to her death from the Acropolis wall, and in some accounts Pandrosus and Herse followed her example.[15] This myth offers a patriotic alternative to the more common story that the Cecropids (or at least one of them) fell to their deaths in terror when they uncovered Erichthonius in his basket. Similar stories of maiden sacrifice or self-sacrifice are told of the daughter of Erechtheus (thrown from the Acropolis by her father in a war against Eleusis) and the daughters of Leos (sacrificed in the same way to save the city from plague).[16]

Jeanmaire finds that in certain African societies such sacrifices have to do with rain-magic.[17] M. Astour too, in interpreting Ugaritic materials, compares the Cecropid myth with West Semitic rites in which sacrificial victims were thrown from a height to promote rain in times of drought.[18] But W. Burkert connects stories of virgin-sacrifice with even earlier strata of human culture: in

----

[15] For testimony see Powell 1906.31 with references. Could this story be an aition for the spring of Aglaurus? See Ninck 1921.11 for the origin of the spring Dirce from the blood of a maiden.

[16] For the daughter of Erechtheus see Apollodorus 3.15.4 and the commentary of Frazer 1921 vol. 2.111; this myth was the subject of Euripides' lost *Erechtheus*, which is quoted in Lycurgus *Contra Leocratem* 100. For the daughters of Leos see Aelius Aristides *Panathenaicus* 119.1 and the discussion in Harrison 1890.lvi-lvii.

[17] Jeanmaire 1939.265.

[18] Astour 1969.17-18.

hunter-warrior societies, according to Burkert, such rituals were carried out before dangerous confrontations--an example of what he terms the "sexualizing of death rites".[19] The water-related names of the Cecropids may strengthen the inference that at some time their sacrifice by "precipitation" was connected with fertilizing moisture, but the mythical context of their sacrifice clearly refers to achieving victory in warfare (a survival in the Greek *polis* of primitive hunter-warrior ritual?). In other aspects of their myth too it is possible that the Cecropids' function shifts from one that is primarily naturalistic and associated with agricultural fertility (or success in the kill) to one that is political, reflecting Athenian civic ideology. The political, patriotic meaning of maiden-sacrifice is certainly its primary meaning when Athenians commemorate their native heroines (e.g. Lycurgus, *Contra Leocratem* 98-101, on the sacrifice of the daughter of Erechtheus).

In other ways too the Cecropids are connected with defenders of Athens. As is well known, the Athenian epheboi, young men in military training, swore their famous oath to defend and preserve their fatherland in the cave of Aglaurus, and Aglaurus was named first in the catalogue of powers to whom they swore. Surely her self-sacrifice presents an ideological model of patriotic loyalty and courage to the young soldiers.[20] Analogies between the Cecropids and the

----

[19] Burkert 1972.70-85 esp. 78-79.

[20] Such is the traditional explanation of Aglaurus' cave as the site of the oath; see e.g. the scholiast on Demosthenes 19.303. See also Merkelbach 1972.

epheboi are further suggested in iconography: a pelike from Camiro depicts the exposure of Erichthonius in typical fashion, except that in place of the terrified girls who have opened the forbidden basket, two figures dressed as epheboi flee from the snakes that have been uncovered along with the infant.[21]

In the ephebic oath instead of *Ágláuros* the variant name *Ágraulos* is attested. The apparent significance of the latter name suggests another and more naturalistic relationship between the epheboi and this Cecropid.[22] The epheboi were especially associated with defense of the borderlands of Attica and with service in the open territory outside the city.[23] Accordingly they were subject to the discomforts and dangers of exposure (closely linked with dewfall, as we have seen, in some military contexts.)[24] Now the adjective *ágr-aulos* 'spending the night in the field' (cf. Latin *ager*) is attested as early as Homer,

----

[21] See Beazley 1963.720 #1. Harrison 1890.xxxii fig. 4 attributes this anomaly to a mistake on the part of the vase painter. In a later discussion of the myth (Harrison 1927.264) she reproduces only part of the pelike, and describes "two sisters" hurrying away, adding (note 3): "The figures on the reverse are actually those of two epheboi, but the vase is almost certainly a copy of some drawing in which Herse and Aglauros are represented." This odd exchange of roles between Cecropids and epheboi confirms the suggestion of Merkelbach 1972 that ephebic devotion to the Cecropids presents in part a negative model, as reinforcement to the epheboi of their duty to maintain awe and respect in the face of religious sanction.

[22] Merkelbach 1972 rightly notes that both variants of the name are significant in Aglaurus'/Agraulus' relationship with the epheboi.

[23] On the subject of the epheboi as youths undergoing an initiatory period of transition, see the celebrated study of Vidal-Naquet 1968.

[24] See the passages from Sophocles and Aeschylus discussed above, pp. 70-73.

and the corresponding noun *agraulîa* is actually used to refer to disagreeable military service in some later texts.[25] Did the epheboi think of Agraulus as a figure who might help or harm them in times when they were living "in the open"? Finally, several Christian writers attest to an ancient cult of "Aglaurus the daughter of Cecrops" at Salamis on Cyprus; here too she is associated with "epheboi", and overtly with the practice of human sacrifice that seems to lie behind her myths on the Acropolis. According to the accounts of her worship (which our sources say was later transferred to Diomedes), the epheboi of Cyprus led a victim three times around her altar before his throat was cut and he was offered as a holocaust.[26]

The complex relationships between Cecropids and epheboi in cult also have a bearing on their relationship to Erichthonius. Although it is not greatly emphasized, in fact the motif of the defense of Athens occurs in several variants of the Erichthonius myth, at its most important junctures. Athena goes to Hephaestus in order to have weapons made, which leads to his attempted rape of her (Apollod. 3.14.1). Again, soon after the birth of Erichthonius she goes to Pellene to get a boulder to fortify the Acropolis; in her absence she entrusts Erichthonius' closed basket to the Cecropids (Antigonus Carystius, *Mir.* 12).[27]

----

[25] E.g. Diod. Sic. 16.15, Dion. Hal. 6.44.

[26] For testimony (three virtually identical descriptions from Porphyry and Eusebius) see Powell 1906.31 and references. See also the discussions in Harrison 1891.354-355 and Nilsson 1957.402.

Erichthonius as an infant of course needs protection, and the three girls to whom he is entrusted have names (and, in the cases of Aglaurus and Pandrosus, also cult places) which suggest their identity on one level as "fertility" figures, connected with the notion of nurturing water. At this point then Erichthonius resembles so many Greek "divine youths": not raised by his own parents (in his case, these would presumably be Earth and Hephaestus), he is put into the care of nymphs or other figures who will raise him, as Jeanmaire has said, "au sein de la nature".[28] But this was not to be the fate of Erichthonius. His nurturer turns out to be not primarily a "nature" divinity but the figure most closely identified with the political and military ideology of Athens, namely Athena herself.

The birth of Erichthonius has important implications for Athenian citizen-soldiers, such as the epheboi must learn to become. Jeanmaire has shown that the mythical infancy of a hero or god often serves as a pattern for male adoles-

----

27 Erichthonius too exhibits some hints of the theme of defense or protection which run through the myth. The snake(s) accompanying him frighten the Cecropids when they succumb to curiosity and open the basket; then Erichthonius is sometimes (e.g. in Pausanias 1.24.7) identified with the snake shown on the statue of Athena in the Parthenon, which is undoubtedly meant to be the "guardian snake of her temple" (scholiast on Aristoph. *Lys.* 758-759). For more testimony on Erichthonius as a snake see Powell 1906.17-19 with references.

28 Jeanmaire 1939.285-290, with examples such as Aeneas and the nymphs of Mt. Ida (*H.Aph.* 256-258) and Aristaeus reared by the Horae and Earth (Pindar, *P.* 9.104-115) or by the Muses and Chiron (Ap. Rhod. 2.510ff.). See also Vidal-Naquet 1972.291-296; and Nilsson 1967.315-317, where Erichthonius specifically is connected with the Minoan *göttliches Kind*, a mortal god attested only in Greek hero stories.

cence and initiation, and Vidal-Naquet has demonstrated that the Athenian epheboi in many ways resemble young initiates into adult male status.[29] Further, it appears that the myth of Erichthonius provides a model for the epheboi at Athens. In a stimulating study of Athenian political identity, N. Loraux shows that the idea of autochthony is of premier importance.[30] The men of Athens (ἄνδρες Ἀθηναῖοι) are united to their city and to one another by their common descent from the *patrîs* 'fatherland', and this is the chief focus of their loyalty. The role of the city in "secular" political ideology, especially in the funeral orations that define the ideal role of the "men of Athens", parallels the role of Athena in the myth. Loraux argues that especially in the fifth-century iconography of the birth of Erichthonius, Athena is shown as his parent and nurturer in all aspects: mother, father and nurse.[31] By extension, because all Athenians are descendants of Erichthonius, Athena is the true parent of them all. The role of nurturer[32] is displaced for an anxious moment onto the Cecropids, three figures whose names may suggest the dangers of exposure as well as

----

29 Jeanmaire 1939.291-292 and *passim*. For the epheboi in this process see Vidal-Naquet 1968; see also Vidal-Naquet 1972.173.

30 Loraux 1979 = Loraux 1981.35-73. So too Bérard 1974.34: "Le sens du mythe d'Erichthonios est d'abord politique."

31 Loraux 1979 esp. 13-17 = Loraux 1981.35-73, esp. 57-65. For the iconography see esp. the important studies of Kron 1976.55-67, Metzger 1976 and Schmidt 1968.

32 For a detailed study of *kourotróphoi* at Athens and throughout the Greek world see Price 1978, esp. 101-107 on Erechtheus and Erichthonius.

the powers of birth and growth, but it quickly reverts to the benevolent, asexual city-protector Athena. And what becomes of the Cecropids, the would-be nurturers of Erichthonius? Once Athena returns and the infant is safe, they disappear from the myth, leaping off the Acropolis either in terror at what they have seen in the basket, for they have violated what must be kept a mystery, or (according to the "patriotic" version of their leap) in ideal dedication to the safety of the city.

The autochthony of Erichthonius then is emblematic of the Athenians' relationship to their city. The epheboi of Athens enter into the fullness of that relationship when they swear their oath of loyalty to the city and its institutions, calling to witness its ancient gods of war and fertility,[33] along with its very trees and crops. In the oath the city is called their "nurturer" (*threpsaménē*) in the fullest ideological sense, as Loraux points out, the citizens are "autochthonous, possessing (the city) as their mother and fatherland" (Lysias, *Fun. Or.* 17).[34] Their entrance into the mysterious cave of Aglaurus/Agraulus under the Acropolis, the site of their oath, may suggest for the epheboi their rebirth into autochthony.[35]

----

33 Siewert 1977.109-110.

34 Quoted in Loraux 1981.66-67.

35 Although for Burkert 1972.77-78 it connotes maiden-sacrifice in preparation for battle; Aglaurus is "the mysteriously killed daughter of the king".

But the birth of Erichthonius is not simple autochthony, for it combines with birth from the native soil the elements of a male progenitor, attempted rape, the protection of the infant by Athena, and the ambivalent role of the Cecropids. As such the myth is unique, even when compared to other myths of autochthony, such as the Spartoi born from the snake's teeth at Thebes, or Typhaon born of Earth at the behest of Hera (*H. Ap.* 331-352). As Peradotto argues, in his critique of Lévi-Strauss' infamous analysis of the Oedipus myth,[36] the nature of human origin presented serious logical problems to the Greeks, and a number of myths address in various ways the difficult question of whether the human race was born from one (i.e. Earth) or born from two (i.e. by bisexual reproduction, which in the beginning must logically have been incestuous). Peradotto offers the suggestion that the birth of Erichthonius presents an excellent "mediation" of the problem: he is born both of Earth and of Hephaestus and Athena.

In addition to the logical problems, it is clear that the question of human origin involved a great deal of anxiety and ambivalance in Greek culture, especially with respect to the functions of male and female parents. As we have seen above,[37] one approach was to posit a complete separation of functions, the male role being to deposit the embryo and the female to serve as vessel or nurse

----

[36] Peradotto 1977, Lévi-Strauss 1955.

[37] See above pp. 24-25.

for the "new-sown fetus" (Aesch. *Eum.* 658-661). In such contexts 'dew' or 'rain', generative moisture from the male Sky, is very much at home, presenting a "cosmological" paradigm for human reproduction. This notion of course underlies many earlier interpretations of the birth of Erichthonius and its relationship to the Arrhephoria, perhaps best represented by A. B. Cook.[38] But this does not explain what happens in the memorable confrontation between Athena and Hephastus!

Greek myths abound in situations where sexual union is *not* desired, and *eérsē* or *drósos* may appear in these contexts too. We have seen this with reference to the daughters of Porthaon (Hes. fr. 26MW) and even to Hippolytus (Eur. *Hipp.* 78), not as an integral part of the myth, to be sure, but as a subtle affirmation of the fact that it is vain for mortals to resist sexuality and procreation, especially when this is contrary to the wishes of gods (Apollo for the daughters of Porthaon, Aphrodite for Hippolytus).[39] In the myth of Erichthonius, however, Athena's rejection of Hephaestus is successful. She maintains her virginal integrity, yet procreation still results. The 'dew' of Hephaestus (his semen is called *eérsē* in Nonnus, *Dion.* 41.64 and a similar image may be hinted at in Callim. *Hec.* fr. 260.19, where the "*drósos* of Hephaestus" seems to

----

[38] E.g. Cook 1940.236-237: Athena and Hephaestus gradually intrude into the parts usually played by Father Sky (i.e. Zeus) and Mother Earth; and 180: the Hersephoroi "were simply conveying the sacred seed of Father Sky into the womb of Mother Earth."

[39] See above pp. 57-58, 64-65.

mean Erichthonius himself)[40] will be generative, if not in union with Athena then when conveyed into Earth. This method of procreation bridges the two "logical" modes of human origin: on the one hand either spontaneous generation from male 'dew'[41] or "pure" autochthony (both being forms of "birth from one"), and on the other hand the mutually welcome union of Sky and Earth, which is fruitful thanks to the 'rain' of Sky (the cosmological model of "birth from two").

The seed of Hephaestus is fecund but a vessel is needed to receive the implanted embryo, and what better one than Earth? Athena's function is now a "mediating" one: she conveys the semen into the ground. Various versions of the myth emphasize this function with picturesque detail, ranging from her simple disappearance at the critical moment (Antigonus Carystius, *Mir.* 12), to wiping the semen from her leg with a tuft of wool and throwing it onto the ground (Apollod. 3.14.6)--or even grinding in it with her foot (Hyg. *Astr.* 2.13).[42] By thus emphatically preserving her virginity, Athena reflects the mysterious power associated with that status in the myths and, no less, the rituals of the city. Following a line of inquiry suggested by I. Chirassi-Colombo, F. Zeitlin has recently studied the ritual status of the virgin girl in Athens.[43] Cit-

----

40 See above pp.23-24.

41 Discussed in detail above, pp. 24-30.

42 The sources are conveniently assembled in Powell 1906.56-59.

ing the catalogue in Aristophanes, *Lys.* 641-647, of the four rituals in which young girls play a major role, Zeitlin concludes: "the ritual status of the virgin girl, from childhood on, is the one which claims the most importance."[44] In the myth of Erichthonius this importance is attributed primarily to Athena, as Zeitlin notes, but I find it suggested as well in the Cecropids, particularly in the figure of Pandrosus, who is usually the innocent sister with regard to opening the basket of Erichthonius. In all versions of the Erichthonius myth, the three girls are virginal figures, daughters of their father with no mention of spouse or children.[45] If Aglaurus/Agraulus typefies the negative aspects of their role, opening the basket and leaping off the Acropolis, then Pandrosus embodies the positive, nurturing, obedient attitude that Athena hoped for when she entrusted the infant to the Cecropids. Pandrosus' function as the nurturing virgin--an adequate replacement for Athena in her absence--is reflected from another standpoint in the fact that the sacred olive tree grows in her sanctuary on the Acropolis. M. Detienne has amply demonstrated the close relationship between this tree and the lives and fortunes of the men of Athens, especially with regard to their double relationship to the city and its cultivated land.[46] More specifically, E. Simon, interpreting the "purpose" of the Arrhephoria, associates it both with

----

[43] Zeitlin 1982.150-153, citing Chirassi-Colombo 1979.47-48.

[44] Zeitlin 1982.150.

[45] Although elsewhere each of the Cecropids is said to be a bride of Hermes: for sources see Powell 1906.66-67 notes 32-39.

[46] Detienne 1970 esp. 10-11.

the nature of the olive crop (for which adequate dew is essential, especially during the dry months after Skiraphorion when the rite was celebrated) and simultaneously with "the life-force of Athens and her people".[47] Pandrosus then may offer not only plant-nurturing moisture from the pure sky, but the youth-nurturing potential of the benevolent female: Erichthonius was delivered up to Athena close by the sacred olive,[48] so that the infant would naturally be associated with Pandrosus' care. But further, this care extends to all citizens, for as Loraux argues, Erichthonius represents the (male) citizens of Athens collectively--especially, I would add, in their period of transition to full adult status. Pandrosus (or Athena Pandrosus) specifically reflects this youth-nurturing aspect of the city goddess. No wonder the epheboi sacrifice to Pandrosus and Athena together with *Gê Kourotróphos* 'youth-nurturing Earth' (*I.G.* II/III^2 1039).

The rite of Arrhephoria offers many close parallels to the myth of Erichthonius. Burkert in particular has noted many details in which the ritual duties of the Arrhephoroi resembles the mythical duties of the Cecropids.[49] Analyzing the duties of the Arrhephoroi--to begin weaving the peplos and to perform the secret nocturnal journey described by Pausanias--Burkert concludes

----

47 Simon 1983.45-46, independently of Detienne.

48 Loraux 1981.41 note 26 cites two Attic red-figured vases (1339 and 1346 in *A.R.V.*) which show the olive tree in scenes of the birth of Erichthonius.

49 Burkert 1966.10-13.

that the Arrhephoria is not so much a rite to renew the fertility of the soil as it is a rite of initiation for the young girls who take part in it. Like many initiates, the girls are separated from their families for an extended period of time, and are given special food and clothing. Most important, according to Burkert, they are introduced to the work of women: weaving and procreation.[50]

This revolutionary rereading of the entire "dew complex" at Athens has elicited an extraordinary amount of discussion. C. Bérard essentially agrees with Burkert's interpretation, and adds from his own point of view that the journey of the Arrhephoroi repeats the descent/ascent movement typical of feminine rites of transition.[51] P. Vidal-Naquet however cautions that the number of girls involved in the rite (there were probably four Arrhephoroi in all, of whom only two made the secret journey) is very small if this is truly a rite of initiation.[52] Independently of Burkert, A. Brelich also considers the Arrhephoroi as female initiates, but he identifies their role as one phase of a complex transition from girlhood to womanhood, which is summarized in the four-part catalogue of *Lysistrata* 641-645 (the same passage cited by Zeitlin, as mentioned above): a young girl would progress from being Arrhephoros to Aletris to Ark-

----

[50] Burkert 1966.13-18.

[51] Bérard 1974.124-125.

[52] Vidal-Naquet 1974.154, although here he mentions only the peplos-weaving. "Initiations by representation" do seem to exist in Greek culture, however: see Bremmer 1978.18.

tos to Kanephoros. As to the small number of "initiates", Brelich's reading of the passage in the *Lysistrata* prompts him to suggest that at least in the fifth century (as opposed to the time of Pausanias, for example), a much larger group may have participated in the four stages of initiation.[53] C. Calame disputes the interpretation of Burkert; he sees the rites of the Arrhephoroi as initiations into the first stages of adolescence rather than the last,[54] and notes especially that the whole area of sexuality is "marked with a negative sign."[55]

All these analyses of the Arrhephoria share the trait of interpretating the ritual from the standpoint of the little girls who participate in it: it is their rite of passage into womanhood or at least a new stage of girlhood.[56] This is akin to interpreting the Erichthonius myth as the story of Pandrosus, Herse and Aglaurus: the girls do have a role in the story, and from their perspective it may indeed be the most interesting role. But it is hardly the point of the story, which focuses primarily on the birth of the first Athenian and his relationship to Athena. Especially because the Arrhephoroi so closely resemble the Cecropids,

----

53 Brelich 1969.229-232.

54 But see Burkert 1966.16-17 for emphasis that the girls' innocence was protected by their explicit ignorance of what they carried in the rite, and by the symbolic nature of the ritual implements and actions.

55 Calame 1977 vol. 1.237-239, summarized also in Zeitlin 1982.151, whose translation I quote here.

56 Kron 1976.68 and Simon 1983.41-46, however, still view the Arrhephoria as a typical fertility rite, with Simon placing special emphasis on its function as a "charm" for the olive crop.

as Burkert has amply demonstrated, it makes sense to try to interpret their rite in terms of the broader perspective that is apparent in the myth: Athenian autochthony. Here I find myself in happy agreement with Zeitlin, who has approached the Arrhephoria from the standpoint of "cultic models of the female" rather than the generative, mediating nature of dew. She puts the matter clearly:

> In the context of the Acropolis, the Arrhephoroi...reenact the autochthonous origins of Erichthonius in order to renew the miraculous "purity" of his birth and to revalidate the virginity of Athena.[57]

In addition, I would add, the rite renews the autochthony of all Athenians as descendants of Erichthonius.

There is one more point where I take issue with most current interpretations of the Arrhephoria, and it has a bearing on the nature of 'dew' both in the myth (i.e. the 'dew' of Hephaestus) and in the rite (interpreted as 'Dew-Bearing'). All the interpretations we have summarized here assume that the terminus of the Arrhephoroi's descent from the top of the Acropolis is a precinct of Aphrodite and Eros on the north slope of the hill.[58] This leads to Burkert's conclusion about the initiates' preparation for "women's work" in procreation, and to the qualifications on that subject expressed by Calame and

----

57 Zeitlin 1982.152.

58 Again I refer the reader to Travlos 1971.228-232 for a summary of the archaeological evidence, description and plates illustrating this fascinating sanctuary.

Zeitlin, who prefer (rightly) to assert the virginal quality of the young Arrhephoroi, who were from seven to eleven years old. The route of the "journey to Aphrodite" depends on how we read a clause in Pausanias 1.27.3:

> ἔστι δὲ περίβολος ἐν τῆι πόλει τῆς καλουμένης ἐν Κήποις Ἀφροδίτης οὐ πόρρω, καὶ δι' αὐτοῦ κάθοδος ὑπόγαιος αὐτομάτη.
>
> There is a precinct on the Acropolis *of / from* the so-called "Aphrodite in the Gardens" *not far away*, and through it a natural underground descent.

A study of Pausanias' language here--our only source for the route of the Arrhephoroi--indicates that their nocturnal journey took them not "to a precinct *of* Aphrodite in the Gardens not far away" (presumably not far away from the Pandroseum, which has just been described); but rather to "a precinct not far away *from* Aphrodite in the Gardens". This is the conclusion reached by E. Kadletz after a survey of the uses of οὐ πόρρω in Pausanias: Kadletz discovers that the adverbial phrase is never used without clear reference to the place "not far away"--and that in 76 of the 82 attestations of the phrase apart from its use here, the place "not far away" is expressed either by a preceding or a following genitive.[59] Assuming that Kadletz's calculations are correct, I find this a persuasive argument that the route of the Arrhephoroi took them only to the cave of Aglaurus, through which there is indeed a passage in the cleft of the rock, and not to the cave of Eros and Aphrodite nearby.[60] The route of the

----

[59] Kadletz 1982.

[60] See Burkert 1966.2 for the usual interpretation; Burkert is aware of the

Arrhephoroi then may be enclosed within the context, or the sexual parameters, of the myth of Erichthonius. They travel from Athena Polias (whose priestess sends them on their mysterious journey) to Aglaurus and back, without being diverted into the precinct of Aphrodite. They are not concerned with the works of Aphrodite, much as they may be concerned with the work of generation. The nature of that generation is suggested by the 'dew' they 'bear', and will be our final consideration.

As Athena effectively conveyed the semen of Hephaestus into the earth to make possible Erichthonius' unique brand of autochthony, so too the Arrhephoroi bear 'dew' in the sense of fecundating moisture.[61] Their rite forms a close parallel to the "artificial insemination" of Earth by Hephaestus, thanks to the mediation of Athena. Furthermore, characteristics of 'dew' in literary contexts may help us to understand more fully its significance for the Arrhephoroi. They too are virginal, daughter figures like Athena daughter of Zeus. We recall that other "daughters of Zeus" dispense 'dew': Ersa (Alcman fr. 57P), the Muses (*Theog.* 81-83), the Charites (Pindar, *I.* 6.63-64, born from Zeus in *Theog.* 907-909). Not only is dew a generative liquid (as seen in its role in spontaneous generation, for example), but it is imagined an effective means of communica-

----

grammatical difficulty in reading οὐ πόρρω but considers the problem solved by Broneer's discovery of the Aphrodite precinct "nearby".

61 *Pace* Martin and Metzger 1976.174, where it is suggested that the Arrhephoroi carry to Aphrodite the balls they played with during their tenure on the Acropolis. See Simon 1983.42 for refutation of this hypothesis.

tion between the world above and the world below--as we have seen in Pindar, *Pythian* 5.96-104, where the dead kings of Cyrene, sprinkled with 'dew' of the victory celebrations, hear about the achievement of their descendant Arcesilas; or in Simonides 125.9-10D, where the 'dew' of grapes growing above the grave is asked to drench the poet buried below.[62]

In what form is their 'dew' carried by the Arrephoroi? If the details of the rite reflect the myth as closely as they seem to, then perhaps they carry down tufts of wool soaked in dew, like the semen-impregnated wool cast onto the earth by Athena (thought to account for the name of *Eri-chthón-ios*, popularly etymologized as a compound of *érion* 'wool' and *chthṓn* 'earth'). As to what they brought back to the Acropolis, I agree with Burkert that the bundle "wrapped up" (ἐγκεκαλυμμένον, Paus. 1.27.3) must have represented a baby,[63] but suspect that it may have been a chthonic, mysterious Erichthonius-like baby, perhaps in the form of a snake.[64] Whatever its form, though, the 'dew' they carry down should be generative, and the way they transmit it should not violate their sexual purity.

----

[62] See above pp. 96-97 for the communicative role of 'dew' in these passages.

[63] Burkert 1966.18.

[64] Cf. the famous scholion on Lucian, *Dial. Meret.* 2.1, which says that in the Thesmophoria, "also called Arrhetophoria," the women carry "images of snakes and figures of male parts."

Finally, the double fate of the Cecropids may also reflect the fate of the Arrhephoroi. To uncover the mysteries would cause severe repercussions, as suggested by the guilty sisters leaping off the Acropolis. To serve the goddess faithfully--at the risk of a dangerous descent of the north edge of the Acropolis at night--leads to their disappearance as Arrhephoroi, for they are replaced by new maidens after their journey (Paus. 1.27.3): does this disappearance perhaps reflect the fate of the heroic sisters leaping off the Acropolis? In any case, for the city to renew its autochthony and its filial relationship to Athena, while maintaining the integrity of its virgin goddess, the Arrhephoroi are brilliantly-chosen intermediaries. Yet the interest of the city is not so much in the young girls who 'carry' the 'dew', but in the function which they perform: renewing its citizens' autochthonous link with the earth under the Acropolis. If this interpretation is valid, then from one perspective the Arrhephoria and the myth of Erichthonius can be seen as a unique local adaptation of the venerable concepts suggested in *drósos* and *eérsē*.

APPENDIXES

I. The Variants of *Eérsē*

It is puzzling that Greek shows some variants with prothetic /e/ (*eérsē* in epic, *éersa* in Pindar), some with prothetic /a/ (*aérsa* glossed as a Cretan form in Hesychius, *aérsē* attested in P. Lit. Lond. 60, of Hellenistic origin), and others without a prothetic vowel (*hérsē* in *Od.* 9.222, Theocritus, etc.; *érsa* in Alcman fr. 57P). This series of variants introduces the question, as yet unresolved, of whether **wers-* actually begins with a laryngeal, and if so, which laryngeal. The laryngeal theory, to be sure, is often helpful in explaining prothetic vowels in Greek.[1] Still, as Cowgill cautions with regard to *eérsē/hérsē*, we should expect a laryngeal in the root to result in prothesis in all the reflexes; moreover, a given laryngeal in the root should generate the same prothesis in its reflexes, i.e. either prothetic /e/ or prothetic /a/, and not sometimes one and sometimes the other.[2] It is to be hoped that further study of laryngeal reflexes in the

----

[1] For the formulation of this theory with regard to **wers-* see Kuryłowicz 1927.104; see Beekes 1969.18-98 for full development and exhaustive treatment of examples; a useful summary may be found in Rix 1976.37, 68-70, concluding (p. 70) that initial laryngeal in the root provides the only way to avoid postulating unconditioned prothetic vowels in Greek. But see Wyatt 1972 for arguments that "the laryngeal theory cannot be taken as a serious explanation for prothesis in Greek" (p. 5). Wyatt himself finds in the series ἀέρσῃ/ἐέρσῃ/ἕρσῃ a classic example of his theory that prothetic /a/ regularly precedes **we/r/C-* in a development restricted to Greek; then /a/ assimilates to /e/, after the prose form ἕρσῃ: cf. Homeric ἀνάεδνον, epic ἔεδνα, prose ἕδνα (Wyatt 1972.37-39).

Greek dialects, or of dialectal variations in prothesis, will provide a more definitive explanation of the variants of *eérsē*.

II. An Etymology of *Drósos*?

Meillet has suggested that *drósos* ultimately derives from **ers-*, the root of Latin *ros* expanded by a prefix /d/ with "purely grammatical significance" and also by a "popular" doubling of /s/ in the root (the latter hypothesis is necessary because otherwise the single intervocalic /s/ would disappear in Greek): *d-ros-s-*.[3] Without reference to this problem, Nagy has compiled a series of pairs in Hittite and Luvian in one instance, Latin and Umbrian in another, in which one member has /s/ and the other has /ss/.[4]

I would cautiously submit that a relationship between *drósos* and **ers-* might also be established along slightly different lines than those suggested by Meillet. If the zero grade of the root (**r̥s-*) were prefixed by a morpheme ending in /n/, such as *pan-* or *en-*, and the epenthetic consonant /d/ were pronounced

----

2 Cowgill 1965.151-153, concluding that the prothetic vowel in ἐέρση does not come from a laryngeal in the root.

3 Meillet 1931.234-6; see also Ernout-Meillet 1959, *s.v. ros*, for a more cautious reappraisal. On the occurrence in Greek of prefixes such as Meillet suggests, see also Schwyzer 1939 vol. 1.417 note 1.

4 Nagy 1974a.71-72; the author warns, however, that he is merely noting the existence of this phenomenon and is not attempting to explain it.

between prefix and original stem, this would produce **pan-d-r̥s-* or *en-d-r̥s-*. In some dialects of Greek--including Mycenaean--according to Wyatt, the reflex of vocalic /r/ in such an environment would be /ro/,[5] resulting in *pandros-* or *endros-*, both of which are attested in Greek: *Pándrosos* (the name of a daughter of Cecrops) and *éndrosos* 'bedewed' (Aesch. *Agam.* 12). The most serious problem with this hypothesis is that epenthetic consonants are normally generated within, and not across, morphological boundaries. This is what we see, for example, in *an-d-ros*, the genitive of *anḗr*, where /d/ is inserted between the /n/ and vocalic /r/ of the stem **anr-*, or in *am-b-rotos*, where epenthetic /b/ comes between /m/ and /r/ in the zero-grade of the stem *mer-* (cf. *mortós* 'mortal' in Hesychius). If an epenthetic consonant really cannot intrude between prefix and stem, then it would be necessary to postulate that the morphological boundaries of *pándrosos* or *éndrosos* first eroded, so that the term was no longer perceived as a compound, but as a simplex meaning 'dewy' or the like. Then the epenthetic consonant was inserted to facilitate pronunciation of the combination *-n-ro-*, and *then* the force of the prefix was felt again and the "prefixed" form was separated along the perceived (but etymologically incorrect) boundaries *pan-drosos* or *en-drosos*. This requires some special pleading, and I would be completely discouraged from suggesting such a development were it not for the somewhat similar development proposed for *brotós* 'mortal' as a back-formation from *ámbrotos*, again along incorrectly perceived morphological bounda-

5 Wyatt 1976.

ries.[6] A final example of possible word-formation from epenthesis rests on the Hesychian gloss which defines *dróps* as *ánthrōpos* 'human being, man'. Is it possible that this peculiar word is a compound of *óps* 'face' and *dr-* derived from *andr-* 'man' (the base of *anér* with epenthetic /d/), along the mistakenly perceived boundary *an-dr-?* If so, this would provide a parallel to the formation of *drósos* from a compound such as *pándrosos* or *éndrosos*.

----

6 This is the process considered most plausible by Chantraine 1968-77, *s.v.* βροτός, although the possibility that **mrotos* itself generated the form βροτός is not discounted.

BIBLIOGRAPHY

Adrados, F. R. 1951. "Sobre las Arreforias o Erreforias." *Emerita* 19.117-183.

Alexiou, M. 1974. *The Ritual Lament in Greek Tradition.* Cambridge.

Arthur, M. 1973. "Early Greece: The Origin of the Western Attitude toward Women." *Arethusa* 6.7-58.

__________. 1982. "Cultural Strategies in Hesiod's *Theogony:* Law, Family, Society." *Arethusa* 15.63-82.

__________. 1983. "The Dream of a World without Women: Poetics and the Circles of Order in the *Theogony* Prooemium." *Arethusa* 16.97-116.

Astour, M. C. 1969. "La triade de déesses de fertilité à Ugarit et en Grèce." *Ugaritica* VI.9-23. Paris.

Austin, N. 1975. *Archery at the Dark of the Moon: Poetic Problems in Homer's Odyssey*. Berkeley, Los Angeles, London.

B = Bergk, T. 1882. *Poetae Lyrici Graeci.*[4] Vol. 1-3. Leipzig.

Beazley, J. D. 1963. *Attic Red Figure Vase Painters.*[2] Oxford.

Bechtel, F. 1917. *Die historischen Personennamen der Griechischen bis zur Kaiserzeit.* Halle. (Reprinted Hildesheim 1964).

Beekes, R. S. P. 1969. *The Development of the Proto-Indo-European Laryngeals in Greek.* The Hague.

Benveniste, E. 1935. *Origines de la formation des noms en indo-européen.* Vol. 1. Paris.

__________. 1948. *Noms d' agent et noms d' action en indo-européen.* (Vol. 2 of Benveniste 1935). Paris.

__________. 1969. *Le vocabulaire des institutions indo-européennes.* Vol. 1-2. Paris.

Bérard, C. 1970. *L' hérôon à la porte de l'Ouest.* (Vol. 3 of *Eretria, fouilles et recherches.*) Berne.

__________. 1974. *Anodoi. Essai sur l'imagerie des passages chthoniens.* Neuchatel.

Beye, C. R. 1974. "Male and Female in the Homeric Poems." *Ramus* 3.87-101.

Bodson, L. 1976. "La stridulation des cigales. Poésie grecque et réalité entomologique." *AC* 45.75-94.

Boedeker, D. D. 1974. *Aphrodite's Entry into Greek Epic.* Leiden.

__________. 1979. "Sappho and Acheron." *Arktouros. Hellenic Studies presented to Bernard M.W. Knox* (Berlin and New York), pp.40-52.

Boisacq, E. 1916. *Dictionnaire étymologique de la langue grecque.* Heidelberg and Paris.

Borthwich, E. K. 1966. "A Grasshopper's Diet. Notes on an Epigram of Meleager and a Fragment of Eubulus." *CQ* 60.103-112.

Brelich, A. 1969. *Paides e parthenoi.* Rome.

Bremer, J. M. 1975. "The Meadow of Love and Two Passages in Euripides' *Hippolytus.*" *Mnemosyne* 28.268-280.

Bremmer, J. 1978. "Heroes, rituals, and the Trojan War." *Studi Storico-Religiosi* 2.5-38.

Broneer, O. 1933. "Eros and Aphrodite on the north slope of the Acropolis." *Hesperia* 2.1, pp. 31-55.

Burkert, W. 1966. "Kekropidensage und Arrhephoria. Vom Initiationsritus zum Panathenäenfest." *Hermes* 94.1-25.

__________. 1972. *Homo Necans: Interpretationen altgriechischer Opferriten und Mythen.* Berlin.

__________. 1977. *Griechischen Religion der archaischen und klassischen Epoche.* Stuttgart, Berlin, Köln, Mainz.

Calame, C. 1977. *Les choeurs de jeunes filles en Grèce archaïque.* Vol. 1-2. Rome.

Carey, C. 1976. "Pindar's eighth Nemean Ode." *Proc. Camb. Phil. Soc.* 202 (n.s. 22). 26-41.

Chantraine, P. 1968-77. *Dictionnaire étymologique de la langue grecque. Histoire des mots.* 5 vols. Paris.

__________. 1968a. *La formation des noms en grec ancien.* Paris.

Chirassi-Colombo, I. 1979. "Paides e Gunaikes: note per una tassonomia del comportamento rituale nella cultura antica." *QUCC* n.s. 1.25-58.

Clarke, W. M. 1974. "The God in the Dew." *L'antiquité classique* 43.57-73.

Cook, A. B. 1940. *Zeus: A Study in Ancient Religion.* Vol. 3. Cambridge.

Cowgill, W. 1965. "Evidence in Greek." In Winter 1965, pp.142-180.

D = Diehl, E., ed. *Anthologia Lyrica.* Vol. 2.[2] Leipzig.

Denniston, J. D. and Page, D. 1957. *Aeschylus Agamemnon.* Oxford.

Detienne, M. 1963. *La notion de daimon dans le pythagorisme ancien.* Paris.

__________. 1970. "L'olivier: un mythe politico-religieux." *RHR* 178.5-23.

__________. 1973. *Les maîtres de vérité dans la Grèce archaïque.*[2] Paris.

__________. 1977. *The Gardens of Adonis: Spices in Greek Mythology.* Atlantic Highlands, N.J. (= Translation of Detienne 1972: *Les Jardins d'Adonis.* Paris.)

Deubner, L. 1932. *Attische Feste.* Berlin.

Dover, K. J., ed. 1968. *Aristophanes Clouds.* Oxford.

Duban, J. M. 1980. "Poets and Kings in the *Theogony* Invocation." *QUCC* n.s. 4.7-21.

duBois, P. 1982. *Centaurs and Amazons. Women and the Prehistory of the Great Chain of Being.* Ann Arbor.

Duchemin, J. 1955. *Pindar: poète et prophète.* Paris.

Dumortier, J. 1935. *Les images dans la poésie d'Eschyle.* Paris.

Dundes, A. 1980. "Wet and Dry, the Evil Eye." In Dundes, ed. *Interpreting Folklore* (Bloomington) pp. 93-133.

Egan, R. B. 1984. "Jerome's Cicada Metaphor (*Ep.* 22.18)." *CW* 77.175-176.

Engelmann, H. and Merkelbach, R. 1973. *Die Inschriften von Erythrai und Klazomenai.* Vol. 2. Bonn.

Ernout, A. and Meillet, A. 1959. *Dictionnaire étymologique de la langue latine.*[4] Paris.

Ferguson, W. S. 1938. "The Salaminioi of Heptaphylai and Sounion." *Hesperia* 7.1-74.

Fick, A. 1910. "Hesychglossen VI." *KZ* 43.130-153.

Finley, J. H., Jr. 1955. *Pindar and Aeschylus.* Cambridge, Mass.

Foley, H. 1978. "Reverse Similes and Sex Roles in the Odyssey." *Arethusa* 11.7-26.

Forbes, K. 1958. "Medial intervocalic -ρσ-, -λσ- in Greek." *Glotta* 36.235-272.

Fraenkel, E. 1950. *Aeschylus Agamemnon.* Vol. 1-3. Oxford.

Frazer, J. G., ed. 1921. *Apollodorus. The Library.* Vol. 1-2. Cambridge (Mass.) and London. (Loeb Classical Library.)

Frisk, H. 1960-70. *Griechisches etymologisches Wörterbuch.* Vol. 1-2. Heidelberg.

Garvie, A.F. 1969. *Aeschylus' Supplices: Play and Trilogy.* Cambridge.

Geldner, K. F. 1951. *Der Rig-Veda aus dem Sanskrit ins Deutsche übersetzt und mit einem laufenden Kommentar verstehen.* Vol. 33-36 in *Harvard Oriental Series.* Cambridge (Mass.), London, Leipzig.

Gianotti, G. F. 1975. *Per una poetica Pindarica.* Torino.

Gilbert, O. 1907. *Die meteorologischen Theorien des griechischen Altertums.* Leipzig.

Giles, P. 1889 "ἕρσαι, πρόγονοι, μέτασσαι." *CR* 3.3-4.

Glenn, J. 1976. "The Phantasies of Phaedra: A Psychoanalytic Reading." *CW* 69.435-442.

Gomme, A. W. 1945. *A Historical Commentary on Thucydides.* Vol. 1: *Introduction and Commentary on Book I.* Oxford.

Graf, F. 1980. "Milch, Honig, und Wein. Zum Verständnis der Libation im gr. Ritual." In *Perennitas: Studi in memoria di Angelo Brelich* (Rome), pp. 209-221.

Hague, R. 1984. "Sappho's Consolation for Atthis, fr. 96LP." *AJP* 105.29-36.

Harrison, J. E. and Verrall, M. 1890. *Mythology and Monuments of Ancient Athens*. London.

Harrison, J. E. 1891. "Mythological Studies I: The Three Daughters of Cecrops." *JHS* 12.350-355.

__________. 1903. *Prolegomena to the Study of Greek Religion*. Cambridge.

__________. 1927. *Themis*.[2] Cambridge.

Henderson, J. 1975. *The Maculate Muse: Obscene Language in Attic Comedy*. New Haven and London.

Herington, C. J. (Forthcoming.) "The Marriage of Earth and Sky in Aeschulus, *Agamemnon* 1388-92."

Hester, D. A. 1966. "A Reply to Professor Georgiev's 'Was stellt die Pelasgertheorie dar?'" *Lingua* 16.274-278.

Hooker, E. 1963. "The Goddess of the Golden Image." *G&R* 10 (suppl.) 17-22.

Irwin, E. 1974. *Colour Terms in Greek Poetry*. Toronto.

Jacoby, F., ed. 1923-58. *Die Fragmenta der griechischen Historiker*. Berlin and Leiden.

Jameson, M. 1965. "Notes on the sacrificial calendar from Erchia." *BCH* 89.154-172.

Jeanmaire, H. 1939. *Couroi et courètes: Essai sur l'éducation spartiate et sur les rites d'adolescence dans l'antiquité classique*. Lille.

Kadletz, E. 1982. "Pausanias 1.27.3 and the Route of the Arrhephoroi." *AJA* 86.445-446.

Kron, U. 1976. *Die zehn attischen Phylenheroen. Geschichte, Mythos, Kult und Darstellung*. Berlin.

Kurtz, D. C. and Boardman, J. 1971. *Greek Burial Customs*. Ithaca, N.Y.

Kuryłowicz, J. 1927. "ə indoeuropéan et *h* hittite." In *Symbolae Grammaticae in honorem Ioannis Rozwadewski* (Cracow), pp. 95-104.

LP = Lobel, E. and Page, D., edd. 1955. *Poetarum Lesbiorum Fragmenta*. Oxford.

LSJ = Liddell, H.G. and Scott, R., edd. Revised by H.S. Jones. 1968. *A Greek-English Lexicon*. Oxford.

Landow, G. P. 1982. *Images of Crisis: Literary Iconology, 1750 to the Present*. Boston, London, Henley.

leGoff, J. and Nora, P., edd. 1974. *Faire de l'histoire*. Vol. 3. Paris.

Lévi-Strauss, C. 1955. "The Structural Study of Myth." In T. A. Sebeok, ed. *Myth: A Symposium* (Bloomington), pp. 81-106.

__________. 1966. *The Savage Mind*. Chicago. (= Translation of Lévi-Strauss 1962. *La pensée sauvage*. Paris.)

Lloyd, G.E.R. 1966. *Polarity and Analogy: Two Types of Argumentation in Early Greek Thought*. Cambridge.

Loraux, N. 1979. "L'autochthonie: une topique athénienne." *Annales E.S.C.* 3-26.

__________. 1981. *Les enfants d'Athéna*. Paris.

Lowenstam, S. 1979. "The Meaning of IE **dhal-*." *TAPA* 109.125-135.

MW = Merkelbach, R. and West, M.L., edd. 1967. *Fragmenta Hesiodea*. Oxford.

Macleod, C. W. 1974. "Two Comparisons in Sappho." *ZPE* 15.217-220.

Martin, R. and Metzger, H. 1976. *La religion grecque*. Paris.

McEvilley, T. 1973. "Sapphic Imagery and Fragment 96." *Hermes* 101.257-278.

Meillet, A. 1931. "Avestique *tkaēša-*." In *Studia indo-iranica: Ehrengabe für Wilhelm Geiger* (Leipzig), pp. 234-236.

Merkelbach, R. 1972. "Aglauros (Die Religion der Epheben)." *ZPE* 9.277-283.

Metzger, H. 1976. "Athéna soulevant de terre le nouveau-né: du geste au mythe." In *Mélanges d'histoire ancienne...P. Collart* (Lausanne), pp. 295-303.

Mikalson, J. D. 1975. *The Sacred and Civil Calendar of the Athenian Year.* Princeton.

Miller, D. G. 1976. "Liquids plus *s* in Ancient Greek." *Glotta* 54.159-172.

Mommsen, A. 1898. *Feste der Stadt Athen in Altertum.* Leipzig.

Motte, A. 1973. *Prairies et jardins de la Grèce Antique.* Académie royale de Belgique. Mémoires de la classe des lettres. Brussels.

Mylonas, G. E. 1961. *Eleusis and the Eleusinian Mysteries.* Princeton.

N = Nauck, A., ed. 1889. *Tragicorum Graecorum Fragmenta.*² Leipzig.

Nagy, G. 1974. *Comparative Studies in Greek and Indic Meter.* Cambridge, Mass.

__________. 1974a. "Six Studies of Sacral Vocabulary Relating to the Fireplace." *HSCP* 78.71-106.

__________. 1979. *The Best of the Achaeans: Concepts of the Hero in Archaic Greek Poetry.* Baltimore and London.

Nilsson, M. P. 1957. *Grieschische Feste von religiöser Bedeutung.* Stuttgart.

__________. 1967. *Geschichte der griechischen Religion.* Vol. 1.³ München.

Ninck, M. 1921. *Die Bedeutung des Wassers im Kult und Leben der Alten.* Leipzig. (Reprinted Darmstadt 1967.)

Norwood, G. 1945. *Pindar.* Berkeley and Los Angeles.

Onians, R. B. 1951. *The Origins of European Thought about the Body, the Mind, the Soul, the World, Time, and Fate.* Cambridge.

Ortner, S.B. 1974. "Is Female to Male as Nature is to Culture?" In Rosaldo and Lamphere 1974, pp. 67-87.

P = Page, D., ed. 1962. *Poetae Melici Graeci.* Oxford.

Parke, H. W. 1977. *Festivals of the Athenians.* Ithaca, N.Y.

Pélékidis, C. 1962. *Histoire de l'éphébie attique des origines à 31 avant Jésus-Christ.* Paris.

Peradotto, J. 1977. "Oedipus and Erichthonius: Some Observations on Paradigmatic and Syntagmatic Order." *Arethusa* 10.85-101.

Petegorsky, D. 1982. *Context and Evocation: Studies in Early Greek and Sanskrit Poetry.* Diss. Berkeley (unpublished).

Pokorny, J. 1959. *Indogermanisches etymologisches Wörterbuch.* Bern and München.

Pope, H. 1953. "Erechtheus and the Erechtheids." In *Studies Presented to David M. Robinson* (St. Louis), pp. 1044-1051.

Powell, B. 1906. *Athenian Mythology: Erichthonius and the Three Daughters of Cecrops.* Ithaca, N.Y. (Reprinted Chicago 1976.)

Price, T. H. 1978. *Kourotrophos. Cults and Representations of the Greek Nursing Deities.* Leiden.

Pucci, P. 1977. *Hesiod and the Language of Poetry.* Baltimore and London.

Puelma, M. 1972. "Sänger und König: Zum Verständnis von Hesiods Tierfabel." *Mus. Helv.* 29.82-109.

Rackham, H. 1967. *Pliny Natural History.* Vol. 3. Cambridge (Mass.) and London. (Loeb Classical Library.)

Rackham, H. 1979. *Pliny Natural History.* Vol. 1.[2] Cambridge (Mass.) and London. (Loeb Classical Library.)

Rankin, A. V. 1974. "Euripides' Hippolytus. A Psychopathological Hero." *Arethusa* 7.71-94.

Rix, H. 1976. *Historische Grammatik des Griechischen: Laut und Formenlehre.* Darmstadt.

Rosaldo, M. Z. and Lamphere, L., edd. 1974. *Woman, Culture, and Society.* Stanford.

Roth, C. P. 1976. "The Kings and the Muses in Hesiod's Theogony." *TAPA* 106.331-338.

Sapir, E. 1939. "The Indo-European Words for 'Tear'." *Language* 15.180-187. (Posthumous notes edited by E. H. Sturtevant.)

Sauvage, A. 1970. "Les insectes dans la poésie romaine." *Latomus* 29.269-296.

Schmidt, M. 1968. "Die Entdeckung des Erichthonios." *Archäologischen Institut, Athenische Abteiling* 83.200-212.

Schwyzer, E. 1939-50. *Griechische Grammatik*. Vol. 1 and 2. München.

Segal, C. P. 1965. "The Tragedy of the *Hippolytus:* The Waters of Ocean and the Untouched Meadow." *HSCP* 70.117-169.

__________. 1967. "Pindar's Seventh Nemean." *TAPA* 98.431-480.

__________. 1967a. "Transition and Ritual in Odysseus' Homecoming." *La parola del passato*. 321-342. (Reprinted in A. Cook, ed. *The Odyssey*. New York 1974, pp. 465-486.)

__________. 1984. "Messages to the Underworld: An Aspect of Poetic Immortalization in Pindar." Forthcoming in *AJP*.

Siewert P. 1977. "The Ephebic Oath in Fifth-Century Athens." *JHS* 97.102-111.

Silk, M. 1974. *Interaction in Poetic Imagery: With Special Reference to Early Greek Poetry*. Cambridge.

Simon, E. 1983. *Festivals of Attica. An Archaeological Commentary*. Madison and London.

Simpson, M. 1969. "The Chariot and the Bow as Metaphors for Poetry in Pindar's Odes." *TAPA* 100.437-473.

Slater, P. E. 1968. *The Glory of Hera. Greek Mythology and the Greek Family*. Boston.

Slater, W. J. 1969. *Lexicon to Pindar*. Berlin.

Snodgrass, R. E. 1967. *Insects, Their Ways and Means of Living*. New York. (Originally published 1930 by Smithsonian Institution Series, Inc.)

Sokolowski, F. 1969. *Lois sacrées des cités grecques*. Paris.

Stanford, W. B., ed. 1959. *The Odyssey of Homer*. Vol. 1-2.[2] London, Melbourne, Toronto, New York.

Stigers, E. S. 1977. "Retreat from the male: Catullus 62 and Sappho's erotic flowers." *Ramus* 6.83-102.

Taylor, C.H., Jr. 1963. "The Obstacles to Odysseus' Return." In C.H. Taylor, Jr., ed. *Essays on the Odyssey* (Bloomington), pp. 87-99.

Thompson, S. 1932-36. *Motif-Index of Folk-Literature*. Bloomington.

Travlos, J. 1971. *A Pictorial Dictionary of Ancient Athens*. New York and Washington.

Usener, H. 1896. *Götternamen. Versuch einer Lehre von der religiösen Begriffsbildung*. Bonn.

Vernant, J. P. 1971. *Mythe et pensée chez les grecs*. Vol. 1-2. Paris.

__________ and Vidal-Naquet, P., edd. 1972. *Mythe et tragédie en Grèce ancienne*. Paris.

__________. 1977. Introduction to Detienne 1977, pp. i-xxxv.

Vidal-Naquet, P. 1968. "Le chausseur noir et l'origine de l'éphébie athénienne." *Annales E.S.C.* 947-964.

__________. 1972. "Le *Philoctète* de Sophocle." In Vernant and Vidal-Naquet 1972. Paris.

__________. 1974. "Les jeunes." in leGoff and Nora 1974.

Wackernagel, J. 1888. "Miscellen zur griechischen Grammatik 12: über die Behandlung von *s* in Verbindung mit *r, l, n, m*." *KZ* 29.124-137. (Reprinted in Wackernagel 1969: *Kleine Schriften*. Vol. 2.627-640. Göttingen.)

Waldron, R. A. 1967. *Sense and Sense Development*. London.

Watkins, C. 1973. "'River' in Celtic and Indo-European.," *Ériu* 24.80-89.

Wendel, C., ed. 1967. *Scholia in Theocritem vetera*. Stuttgart.

West, M. L. 1966. *Hesiod Theogony*. Oxford.

__________. 1978. *Hesiod Works and Days*. Oxford.

Whallon, W. 1969. *Formula, Character, and Context: Studies in Homeric, Old English, and Old Testament Poetry*. Cambridge, Mass.

White, K. D. 1970. *Roman Farming*. Ithaca, N.Y.

Wilamowitz-Moellendorff, U. von. 1926. *Euripides Ion*. Berlin.

__________. 1959. *Der Glaube der Hellenen*. Vol. 1.[3] Darmstadt.

Windekens, A. J. van. 1956. "Gr. δρόσος, '*Tau*'." *KZ* 73.26-27.

Winter, W., ed. 1965. *Evidence for Laryngeals*. London, The Hague, Paris.

Wyatt, W. F. 1972. *The Greek Prothetic Vowel*. Cleveland.

__________. 1976. "Sonant /r/ and Greek Dialectology." *Studi Micenei ed egeoanatolici* 13.106.122.

Young, D. C. 1968. *Three Odes of Pindar*. Leiden.

Zeitlin, F. 1978. "The Dynamics of Misogyny: Myth and Myth-making in the *Oresteia*." *Arethusa* 11.149-184.

__________. 1982. "Cultic Models of the Female: Rites of Dionysus and Demeter." *Arethusa* 15.129-157.

INDEX OF PASSAGES